Designed to Evolve

Designed to Evolve

Discovering God through Modern Science

By Christopher S. Davis

MOTIF PRESS

MOTIF PRESS

Published in the United States of America by Motif Press.

Copyright © 2015
Christopher S. Davis
First Print Edition 2015

Table of Contents

Preface

This book is written for those who wonder if there is a logical basis for belief in God in the modern age. For this reason, this book takes a very critical look into the evidence and counter-evidence to provide an answer. Through this, all realms of science will be scoured: cosmology, particle physics, quantum mechanics, geology, and natural history as revealed by the fossil record. Many aspects of biology will also be examined, including a deep dive into evolution. While the undecided may appreciate the critical methods in questioning God's existence, this may be disconcerting to a firm believer. This investigation proceeds under the assumption that basic science is valid, and it is from science that the question of God is tested. It is not because this is the only avenue, or even the best avenue, but because this is the route that many who are still undecided will recognize as a legitimate test. *Designed to Evolve* takes you on a journey through time and space, piecing together clues from mainstream science that provide extensive evidence for a created universe.

Introduction

"In the beginning God created the heavens and the earth," reads the opening line of six billion bibles on this planet. Penned over 3000 years ago, this profound claim has been accepted for millennia. In the fourth century BC, Aristotle presumed that there must be a single supreme creator to explain the existence of the natural order. Though he did not know about the God of the Bible, he reasoned that this creator would be unchanging, immortal, perfect, and good.[1] This was in sharp contrast to the mythology of his culture, where there were many gods taking on human behaviors of treachery and deceit.

Are these ancient claims for a creator valid in the twenty-first century? Has science supplanted philosophy and theology in explaining our existence? What can science tell us about the existence of God?

In the media and in our classrooms, we hear the increasingly common message that science now explains our existence through natural processes alone. From cosmology, we have learned that the universe began from a single explosive event billions of years ago, known as the Big Bang. The Earth itself is some 4.5 billion years old, having life present on it most of this time. Evolution claims to explain not only our origin but our behaviors as well.

Yet billions of people in the modern age still believe in God. When we examine the scriptures that have shaped our culture, a variety of perspectives can be found on how to relate them to modern science. Some religious viewpoints completely disregard the findings of science as it relates to age, postulating that our world is merely thousands of years old. But these views often originate from those without an understanding of astrophysics, geology, or biology. It is no wonder that such conclusions draw contempt from those who are trained in the various sciences. However, there are some theologians who are more receptive to science, viewing it as a tool to shape our understanding of creation alongside scripture.

For many it is not science that is rejected but religion. Atheism is on the rise throughout the world, and there is extensive support in the scientific community for this position. Is our place in the universe too perfect to be an accident? The naturalist would say, "No, we are evolved to suit our environment. Over the eons, among billions of worlds, there exists one that happens to be just right for us." The naturalist would dismiss scriptures as myth and belief as superstition.

Are we here by chance, as the culmination of millions of random events, on vast scales of time, or were we created for a purpose? Does any physical evidence exist, that can be detected by science, to support the popular belief in a creator? The purpose of this book is to address these questions with the aid of modern science.

This pursuit is not just academic, but a very personal journey. Like many who grew up in the faith, as an adult I had to come to my own conclusions. My education was heavy in math and science, and I actually loved it. Science is so exciting; it reveals the inner-workings of the universe. But science has a dark side, or so it would seem. From it come claims that seem to strike at the heart of many religious teachings.

For years I wrestled with two world views, using each one in separate compartments of my life. For I believed in God, as the creator of the world. And I lived and worked in science, having a firm understanding of the physical laws that govern reality. Occasionally, I would become tempted by "creation science" theories that promised to demonstrate how our Earth could be very young. But careful review quickly led to the realization that the details of these ideas were scientifically flawed. So it seemed convenient to simply avoid any subject matter that would cause my world views to collide. In time, this ostrich with its head in the sand could not quench the desire for truth.

So I decided to embark on a quest for knowledge. I would temporarily shelve my opinions, my beliefs, my prior education, and all that I had come to understand about the world. This was not a rejection of faith but a step of faith. For I knew the truth could be found, if only I put forth the effort to find it.

For years I have researched every aspect of origins and how our universe, our planet, and our own species came to be. Examining the claims of many scientists, I accept nothing without proof. The solution to the mystery of existence is recorded in the elements around us. From the nearest stars, to distant galaxies, to the afterglow of the universe's birth, photons reveal the universe's history in exquisite detail. The rocky layers within the Earth have much to say about life's history, through the fossils within them. Even our own DNA is imprinted with evidence about the nature of our origin. This odyssey has led to fascinating discoveries about the reality in which we live, revealed by scientists of the present age. Is our existence due to intentional creation or a statistical fluke? This book analyzes both viewpoints, and shows why the evidence favors reason and purpose in the universe. Through this experience, I faced the flaws in my own world view and emerged with a deeper understanding of reality. By studying creation, the nature of God is also revealed.

I invite you to take this journey, while we explore the origin of the universe, our planet, and life as determined by modern science. We will examine the evidence, biblical predictions, claims of atheists and theologians, and many modern discoveries. This can be quite disheartening as the facts are laid out, and the hard questions are asked. Without regard to any opinion, the claims of both sides are scrutinized. Then decide for yourself as the mystery unfolds. Some believe that existence on every level can be attributed to chance, but the case for luck does not hold up to modern findings. A complete view of reality emerges. This is a dangerous venture that may disrupt your own world view. But it is worth the effort. To discover the secrets of creation is to detect the fingerprints of God.

Chapter One

The State of the Universe

To look up at the night sky, far from city lights, yields an impressive view of our galaxy, the Milky Way. This is viewed as a thick band of stars and dust stretching from one end of the sky to the opposite horizon. As beautiful as this is, it is nothing like what is seen with the deeper views of a large telescope.

What can be seen of our universe is nothing less than astounding. In every direction, over every patch of sky, we find the same thing: galaxies. Most galaxies reside in clusters or in super clusters, some of which contain millions of galaxies. The visible universe contains, on the order of, a trillion or more galaxies. They tend to clump together in sheets and filaments forming a cosmic web of matter interlaced around immense voids. Though each galaxy is unique, the distribution of galaxies averaged over the largest scales is the same in all directions. That is to say that the average density is constant in all directions of space.

A typical galaxy will contain hundreds of billions or up to a trillion stars. It is believed that a super-massive black hole resides at the center of most, if not all galaxies. The black hole in the center of our galaxy weighs in at three million times the mass of our sun. Our galaxy has a spiral disk shape with a large bulge at the center. The central stars have randomly oriented elliptical orbits while the stars in the disk have nearly circular orbits about the galactic center. Our galaxy is located in the Local Group, a small group of about forty galaxies. This cluster is gravitationally bound to the Virgo Super Cluster that is comprised of over a million galaxies of its own.

Our sun is a fairly typical star of intermediate size with eight planets and several dwarf planets (one of which is Pluto). Other dwarf planets include Ceres (in the asteroid belt), and bodies more distant than Pluto (Haumea, Mauki Mauki, and Eris). Though most of the stars visible in the night sky are much larger than the sun, the majority of stars throughout the galaxy are smaller. Since larger stars are brighter they can be seen easily from a great distance.

Up to a few years ago, we had no clear picture of the prevalence of planetary systems like the one in which we live. The Kepler spacecraft has changed this. Designed not only to find planets around other stars, Kepler has provided an estimation of the frequency for each type of planet. Analysis of the first three years of data has resulted in the discovery of more than 3500 potential planets, including some of the most Earth-like planets ever found. Though truly Earth-like planets are absent from the list, this could change in the future as planets with larger orbits are discovered. As of this writing, Kepler has found 674 planet candidates that are roughly Earth's size.[1] Some of these were within the habitable zone of their star. Twenty percent of the planets were part of multiple-planet systems, with one star system found to have at least six planets.[2]

Kepler uses the transit method to detect planets. With this method, transits are recorded when a planet passes in front of the host star during the course of its orbit. This only works for planetary systems whose orbital plane is lined up so that we see it edge on, or for about two percent of nearby extrasolar planets. Because planets that are close to their host star have shorter orbital periods, they are found first. As the remaining Kepler data was processed, researchers were able to find planets that were farther out. Though the Kepler primary mission has ended due to mechanical difficulties, future missions with greater sensitivity may one day find Earth-like planets in Earth-like orbits around sun-like stars.

With many discoveries yet to be made, Kepler has already demonstrated one simple fact. Planets are quite common in the universe. By extrapolating the Kepler results, some scientists have estimated that there are billions of terrestrial planets in our galaxy. This fact seems to raise the odds of one day finding another planet with life. Kepler has also demonstrated the wide variety of planetary systems, most of which are very different from our own. Many solar systems have planets crowding very close to their star. Double star systems contain two suns, and some of these have been found to possess planets as well. Parameters like star type, planet size, orbital configuration, and chemical composition, are known to vary widely between solar systems. Our place in the cosmos seems to shrink as we consider all the worlds that must occupy our galaxy and even more within the entire universe.

Besides the sheer number of celestial lands, the distances between them are vast. Light travels at 300,000 km/s (186,000 mi/s). So it takes eight minutes for light from the sun to traverse 150 million kilometers (93 million miles) to reach Earth.[3] It takes an hour to reach Jupiter and six hours to reach Pluto. This is to say that Pluto is six light-hours away from the sun. Alpha Centari is the nearest star to the sun, yet light from it takes 4.3 years to make the trip. This is the closest of the approximately 200 billion stars in our galaxy. The disk of the Milky Way is about 100,000 light-years across,[4] and stars in the halo and in globular clusters are even farther out. The sun is 24,000 light-years from the galactic center.[5] On a good summer night, far from city lights, you can see the nearest major galaxy, Andromeda, with the naked eye. A better view is obtained with a pair of binoculars. Though you can see the luminous central bulge, it appears very faint from here, about 2.5 million light-years away.[6]

When we view the most distant galaxies, we observe them as they existed billions of years ago when the light we now see was emitted. These galaxies are oddly shaped and are smaller than typical in our own galactic neighborhood. They are young and numerous and give us a glimpse of the universe in its infancy.

The universe hosts billions of galaxies, each containing billions of stars capable of possessing planets. It might seem that we are simply lost in the vastness of space among innumerable desolate worlds that are lifeless and void. Barren rocks are the norm; it is our world that is truly strange.

As we survey the heavens, what we see leads us to consider many questions. Where did it all come from? Why is our planet so well suited for us? How did life come to exist on Earth? Does life exist anywhere beyond our planet? Over the centuries, many discoveries have added to our knowledge of science. Over time we have began to address some of these enigmas, giving us a better picture of the deep history of the universe.

History of Discovery

The prevailing theory of the universe prior to 1927 was the steady state theory. It was believed that the universe was eternal, having always been essentially the same as it is now and always would be. This theory was in support of the atheist's view, given that an eternal universe does not need to be created. So it seemed that science could provide a naturalistic explanation for the universe's existence without the need for God.

In 1905, Einstein proposed his theory of special relativity. This included the famous formula "$E=mc^2$" as well as notions of a constant speed of light, velocity being relative, and time dilation. This demonstrated that time was a local property and part of the universe itself.

In 1916, Einstein developed the theory of general relativity.[7] Through this theory, gravity is described as the result of curvature in space-time. This also included Einstein's field equations. These equations describe the distortion of space-time, that is caused by gravity, in mathematical terms. Space was now determined to be a local property as well. This forced theorists to recognize that space and time are not immutable, but are formable and changeable aspects of the universe.

It was from Einstein's field equations that priest and physicist Abbe Georges Lemaitre derived solutions to these equations and presented the Big Bang model of the universe in 1927.[8] According to his theory, the universe began from a singularity of infinite density, then expanded rapidly to a great size and continues to expand to this day. The early stages of the universe would have been at incredible temperatures. Initially, only photons and exotic particles would be able to exist. As the universe expanded, it would cool. Eventually temperatures would be low enough for normal matter to exist. As time progressed, gravitational attraction would cause the initial gases to coalesce into clouds, stars, and galaxies. His theory received mixed reviews at first. Many objected to the idea that the universe had a beginning on philosophical grounds; it sounded too much like a creation theory. If the universe had a beginning then it needs an explanation for its origin. It would be decades before enough evidence would be found to allow for the full support of the scientific community.

In 1929 Edwin Hubble measured distances for a large number of galaxies and correlated this with their velocities.[9] He discovered that the greater the distance to a galaxy from Earth, the faster that galaxy was moving away from Earth. This was true for all distant galaxies, in all directions. The relationship between distance(D) and velocity(V) was found by Hubble's Law (V=HD).[10] Where H is

Hubble's constant (68.1 km/s/MPc).[11] This discovery provided the first evidence in support of the Big Bang, by showing that we live in an expanding universe. This finding itself was enough to falsify the steady state theory of the universe. But Big Bang theory was still hotly debated for many years after.

Another prediction of the Big Bang model was that residual radiation should be detectable today from the Big Bang.[12] This would occur when the primordial elements of the Big Bang cooled enough to deionize so that light could pass freely though the neutral gases in space. It was not until 1965, that the cosmic microwave background (CMB) radiation was accidentally discovered while doing satellite communications experiments at Bell Labs. The CMB persisted in all directions and with a homogeneous spectrum corresponding to a black body temperature of 2.7 Kelvin (-453.0 °F).[13] No other explanation can adequately explain the CMB, outside of the Big Bang model. The present age of the universe can be calculated from CMB data. Scientists know it would have been emitted at 3000 K (4942 °F), the temperature that the neutral hydrogen gas became stable. Dividing this by the current CMB temperature gives us a result for the amount of expansion that has occurred since the CMB was emitted. This is an expansion of about 1100 times.[14] The effective rate of expansion is also determined from the CMB. With this information, the age of the universe is estimated at 13.8 billion years.[15]

By this point the debates were ended, and most scientists now accept Big Bang theory as the best description of the universe's origin. It was also the nail in the coffin for the steady state and other competing theories of the universe. Big bang theory shows us that the universe is not eternal into the past; it has a beginning. But the success of this theory does not end there, the many stages of the universe's evolution can be calculated as well.

The Expanding Universe

The Big Bang theory proposes that the universe began out of a single explosive event. At that instant the universe began microscopic in size and grew rapidly to galactic proportions. All the energy that exists now has existed since the beginning and has remained constant throughout history. This was not an explosion expanding within a vast void of empty space; it was space itself that was expanding. All existing space has been completely occupied by this cosmic fire ball at all times.

To understand how space can expand, let's look at the simplest geometry the universe could have: a closed volume with positive curvature. As a two dimensional analogy, imagine that space exists as the surface of a small balloon. Someone in a tiny spaceship can travel in a straight line along the balloon's surface, but no matter which direction they travel, eventually they come back to their point of origin. To the space traveler, his path was perfectly straight. But because space itself is curved, it is not infinite in any direction, the whole of space is finite in volume.

At an early time, this balloon is very small and points are marked to represent galaxies all over its surface. As the balloon is filled, its surface area increases, and the points we marked become farther apart. The filling continues in a way that causes a constant increase in distance over each inch of length per second. So an observer at one point on the surface of the balloon would observe all other points to be receding away. When he measures the velocity of various points, he would find that the more distant points would be moving away faster than nearer points. However, observers at all points would consider themselves to be at rest and observe the same recession for distant objects. Since the balloon's surface area is increasing, any matter distributed on it will become more dispersed over time. (Or the average density will decrease.) After much time, the balloon would become so large that an observer on it, would have great difficulty measuring any curvature, because it would be nearly flat in the small region that he could see and measure. Similarly, it is difficult to perceive the curvature of the Earth when standing on its surface.

In the same way, the universe began microscopic in size and expanded over time to its current dimensions. It is apparent from this analogy, how space could be finite in volume yet have no edge or boundary. We can also see how space itself can increase in volume over time. But the universe is not two dimensional like the surface of the balloon; there are three spacial dimensions. This implies that a fourth spacial dimension is required to curve space through, but the curvature is so slight that we can not currently detect it. It is therefore unknown if the universe is finite or infinite in size.

Here is an overview of the various stages of Big Bang cosmology, describing the universe's beginning and subsequent evolution throughout time.

The Planck Epoch

This is the instant of creation. It is not known how or why it began. It can only be observed that it did begin. It makes little sense to ask what happened before, because time itself began at this instant and there was no time before (time is a property of the universe.) The Big Bang does not describe matter expanding to fill empty space. Space itself is what is expanding. All of space, at this moment was extremely small (smaller than the nucleus of an atom) and filled with a dense energy field.

Because the energy of the universe has been constant since the beginning, the universe was at an unimaginably high energy density. Normal matter or even more familiar subatomic particles could not have existed during this scorching moment.

Inflation

Inflation theory is a modification to the original Big Bang theory, but it is now widely accepted as part of the standard Big Bang model. This theory postulates that the universe underwent a brief period of rapid expansion, soon after it was born, expanding at an exponential rate. After about 10^{-32} of a second of inflation, the part of the universe that we can now see could have been held in your hand. However, it was extremely dense, containing within it all the mass and energy of the visible universe.

It is believed that this period of inflation is what caused the universe to become extremely uniform as it would have stretched small isothermal regions beyond the cosmic horizon. This would also have caused the universe to be extremely flat on scales that can be measured.[16] All of this rapid expansion could have made the universe too smooth for things like stars and galaxies to form. But quantum fluctuations near the end of inflation would have been stretched to macroscopic scales.[17] These slight density variations would provide the seeds of galaxy formation in the eons to come.

Baryogenesis

The universe continues to expand at a slower rate after inflation ends. As it expands it cools, by the same laws of thermodynamics that exist today. The conditions at the Planck epoch and during inflation are quite speculative, but the physics of this era are well explored in particle accelerators of our time. Once the temperature is low enough, protons and neutrons become stable. Neutrons form at the rate of about one for every seven protons, resulting from the value of the Boltzmann factor ($e^{(m_n - m_p)c^2/kT}$).[18][19] Electrons and other subatomic particles would have already formed by this time. A dense soup of subatomic particles filled the universe.

Nucleosynthesis

After continued expansion, the protons and neutrons cool (slow down) enough to form atomic nuclei (76% hydrogen, 24% helium, and very trace amounts of lithium and beryllium.[20] These are the first four elements of the periodic table.) The proportion of these elements predicted from theory agrees closely with the observed relative abundances of the elements in the universe. This phase begins between 100 and 300 seconds after the Planck epoch and lasts only a few minutes. The nuclei exist as ions, since they are too hot to combine with electrons. Light is continually absorbed and re-emitted by the ions, making the hot plasma opaque.

Decoupling

After about 380,000 years the universe cools to 3000 Kelvin (4942 °F). At this temperature, neutral hydrogen and helium atoms are stable. Once neutral atoms form, light passes through the gas freely. This is the time when the cosmic microwave background (CMB) was emitted.[21] It has continued to travel through space until this day. This faint background radiation is still detectable as the afterglow of the Big Bang.

Galaxy Formation

Slightly denser regions of gas begin to collapse under the pull of gravity. These gas clouds form small galaxy systems from which the first stars are born. The universe would be about four hundred million years old before these new lights begin to shine. Because only hydrogen and helium exist, the first stars were believed to be unusually massive.[22] They burned hot and fast fusing hydrogen into successively heavier elements. A very large star will continue to fuse elements until it reaches iron, which collects in the stellar core. Once the iron core reaches 1.4 solar masses, the star will go supernova, creating all of the natural elements heavier than iron.[23] The blast from the supernova spreads the newly manufactured elements into the galactic medium. These elements are then available for the formation of future generations of stars and planets.

Coalescence

Many generations of stars are born and die. Further enriching the universe with heavy elements. Since the beginning until now, stars have converted about 2% of the universe's original hydrogen into other elements.[24] Small galaxies collide and coalesce into larger galaxies. Groups of galaxies are drawn together by gravity to form clusters and super-clusters of galaxies (thousands or millions of galaxies). All the while, the universe is still expanding, so these massive clusters are getting farther away from each other. This process continues today (13.8 billion years after the Planck epoch).

Continued Success

As we study the universe, Big Bang theory is verified all the more. Particle physics is able to predict the expected abundances of the elements. The Hubble Space Telescope reveals the beginnings of galaxy formation at great distances, showing rapid star bursts, frequent mergers, and other predictions of the early universe. The CMB has been mapped across the entire sky; its pattern of variation is well aligned with expectations from theory. There is little doubt in Big Bang theory these days; it's the initial stages (the first second) and the minor features that are debated now.

The great success of this priest's theory has continued to find support in science and compels us to consider what caused the Big Bang to occur. Big Bang theory describes the universe's evolution after it began, but says nothing about what initiated it. A closer look will be necessary to determine if there are any indications as to what actually caused the Big Bang. General relativity shows us the ancient past, and it also tells us about the future. As we look into the future of the universe, general relativity can help us evaluate the merits of each possibility.

Chapter Two

The Fate of the Universe

People have often contemplated how the Earth might end. We may destroy it ourselves through weapons or some other technological blunder. It is possible that a large asteroid or a planet whose orbit has shifted could collide with the Earth. Or a nearby supernova could sterilize our planet, though the bulk of it would remain intact. But if none of these fates ever decimate our planet, it will likely be engulfed by the sun when it becomes a red giant towards the end of its life, sometime in the next few billion years. As disheartening as it is to consider the end of our world, there is an even greater end approaching: *the end of the universe*.

Entropy: The Dispersion of Energy

Fundamental to all physics are the first two laws of thermodynamics. The first law is simple: energy can never be created nor destroyed. It can change forms, such as from kinetic to thermal, but none of the energy is lost. This leads to the second law: entropy always increases. Entropy is the measure of randomness or the dispersion of energy. That is, energy must always flow to a more dispersed state.

These two laws forbid perpetual motion machines and the like. They have been known for centuries and are crucial to nearly every branch of science and engineering. You can see these laws at work with a freshly-brewed cup of coffee. It starts with a concentrated store of energy in the form of heat at an elevated temperature. Steam rises off, filling the cool room with its scent. Small eddies swirl in the cup as cooling liquid at the surface sinks to the bottom and warmer fluids rise to the top. But in time, the cup is cold, the coffee is still, and the room is very slightly warmer than it was before. The energy in the coffee is not lost but has spread out and is no longer useful. This is entropy as described by the Second Law of Thermodynamics. The greater the entropy, the less useful the energy becomes. As energy disperses, its usefulness is lost.

You cannot wait for the energy to return to the coffee again so that you can finally take a drink. It will not. This is the second law at work. The only way to decrease the entropy of a system is to increase it somewhere else by a greater amount. You can put the coffee into the microwave and reheat it, but this comes at the expense of the low entropy state of electricity being converted to heat at higher entropy.

A way of visualizing entropy comes from statistics. Entropy must increase because the number of possible high entropy states is much greater than the number of low entropy states. For example, our cold cup of coffee sits on the table. It is still an orderly system in its geometric state, so its entropy can still be increased. There are only a few ways for the fluid to be arranged where it looks the same, contained in the cup. The cup must remain upright in all such states. However, if the cup is tipped, the coffee will flow, making use of gravitational energy. It will run in all directions across the table, then drip from the edges to the floor. The coffee now has many possible arrangements it can be in. Each drop could potentially occupy any number of positions throughout the room. It is the larger number of possible configurations of the end state that dictate the progression to a higher entropy state whenever an energy transfer occurs. Because the second law is due to probability, it is not simply a discovered property of physics, but a purely mathematical result. This is why it is called a law; it is absolute, as sure as two plus two equals four.

Energy, Entropy, and the Heat Death of the Universe

No closed system can experience a decrease in entropy. The universe is the ultimate closed system. All the energy that exists now has existed since the beginning. And since the beginning entropy has been increasing. Stars have been converting mass energy stored in hydrogen gas into helium, light, and heat for billions of years. Many generations of stars have already lived and died in the history of the universe. When a star uses up all its hydrogen, it may consume other elements if it has sufficient mass, but eventually its fuel runs out. The star fades and contracts into a dense cooling ember. A large star will end as a supernova, blowing most of the star's outer layers into space. The star's core may collapse into either a neutron star or a black hole.

It was previously mentioned that stars have already consumed two percent of the universe's original hydrogen.[1] Eventually stars could consume all the universe's hydrogen (and in very large stars, all elements up to iron). After this, star mergers or stars falling into black holes could provide a brief supply of energy. But eventually all nearby stars will merge, leaving only black dwarfs (cool stellar remnants of small stars), neutron stars, and black holes. The following chart shows how long a star can maintain fusion in its core and the type of star that results after its fuel is consumed. Over billions of years, white dwarfs slowly cool and fade into black dwarfs as left over heat dissipates into space.

Lifespan of Stars by Mass

Size*	Post-Fusion Phase**	Lifespan***
0.08 $M_\odot$	He White Dwarf	11 trillion years
0.2 $M_\odot$	He White Dwarf	1 trillion years
0.5 $M_\odot$	C-O White Dwarf	20 billion years
1 $M_\odot$	C-O White Dwarf	10 billion years
2 $M_\odot$	C-O White Dwarf	800 million years
10 $M_\odot$	O-Ne-Mg White Dwarf	20 million years
20 $M_\odot$	Supernova, Neutron Star	10 million years
40 $M_\odot$	Supernova, Black Hole	1 million years

* Size relative to the sun's mass (1 $M_\odot$)
** The type of star that results after fusion has ceased (star death). This remnant will slowly cool and fade.
*** Length of time that fusion is sustained.

234

With no useful energy left, the universe will grow cold and dark, leaving no energy available for stars to shine or for life. This is known as a heat death. However, this heat death could only occur if the universe lasts long enough (about a hundred trillion years).[5] But other dangers exist that could bring about an early end. So this is really a best case scenario.

The Impact of General Relativity

Einstein's famous field equations of general relativity describe how space and time are distorted by the presence of gravity. Indeed gravity acts against the initial momentum of the universe's expansion. The strength of gravity's influence on the universe's expansion rate is directly related to the average density of the universe. This results in three possible outcomes. If the density is high enough, gravity could eventually overwhelm the universe's expansion, slowing it over time, and eventually bring it to a halt. Once expansion has stopped, there would be nothing to stop gravity from pulling it all into reverse, causing the universe to contract. As the universe contracts, the density would increase, thereby increasing the effects of gravity and the rate of contraction as well. The universe would continue to contract at an increasing rate until it would end in much the same way as it began, as a singularity, with all the matter in the universe collapsing into a single point of infinite density. This is known as the Big Crunch.[6] The universe would spend half of its life in a decelerating expansionary state and the other half in a state of accelerating contraction. This model represents the universe with positively curved space-time, with space returning to zero volume in a finite amount of time.

Now if the density of the universe is equal to the density required for gravity to just barely halt the expansion, only after infinite time, then the universe's geometry is said to be flat. The density required for this to occur is called the critical density. In such a case the universe would continue to expand forever at an ever slowing rate.[7] Though the universe and the dispersed energy within it would continue indefinitely, the universe would suffer the previously described heat death, at which point, life in the universe would no longer be possible.

If the energy density is less than the amount required for gravity to halt the universe's expansion, then the universe has negatively curved space-time.[8] In this case the end may be much like the scenario for a flat space-time, a heat death. But the expansion will continue forever at a nearly constant rate. Distant galaxies would eventually disappear from view as the distance between us increases faster than light can make the trip. In the later stages, long before the heat death, all galaxies that are not gravitationally bound would be separated by such vast and increasing distances that they would be undetectable to the rest of the universe.

Accelerating Expansion and the Cosmological Constant

Now it gets a little more complicated. There is another term in Einstein's equation theorized to exist, the cosmological constant. Einstein originally added this parameter to balance the force of gravity and allow for a steady state universe, which neither expands nor contracts. But when it was discovered that the universe was in fact expanding, it seemed to not be needed. So he removed it, recognizing that the steady state universe would be unstable even with the parameter. He called it the biggest blunder of his career, since he missed being able to predict the expanding universe because of it. But recent evidence suggests that even though the universe is expanding, the cosmological constant may actually be required in his equations after all.

In 1998, the Supernova Cosmology Project, which had begun expecting to measure the deceleration of the universe, instead discovered that the universe's expansion was actually accelerating.[9] Since the energy density of the universe has been found to be very near the critical density, this acceleration is believed to be due to a positive value for the cosmological constant. This is where dark energy comes in. We do not know what dark energy is, but we can see its effect on the universe's expansion.

Because we do not fully understand dark energy, it is difficult to predict its future impact on the cosmological constant. So far evidence suggests it is indeed constant.[10] So as gravity is weakened by the ever-expanding cosmos, the cosmological constant remains in full force, increasing the rate of expansion, so that even if the universe is flat or even has slight positive curvature, it will continue to expand forever.

However if the cosmological constant increases with time, as some have suggested, then the universe could undergo a runaway expansion called the Big Rip. In this event, the cosmic expansion would continue at an increasing rate until it becomes infinite. In this scenario, galaxy clusters would become isolated as the rest of the universe expands out of view. Then galaxies would become separated from the clusters as the expansion begins to overtake the force of gravity at large scales. Only stars in our home galaxy would be visible in the night sky. Eventually though, the expansion would become so rapid that stars would be stripped from their host galaxies. In the last epoch, the night sky would be dark, since all other stars would be too far away to be seen. Then even planets would be pulled from their home stars. Finally, solid bodies and then even atoms would be ripped apart as even the nuclear forces become nullified by the unrelenting expansion.

Now the first end doesn't seem as bad when we consider the results for a positively curved space-time or for an increasing cosmological constant. This is because both of these would result in a much shorter lifespan of the universe (and a more violent end) than any scenario that leads to a heat death as a result of the Second Law of Thermodynamics.

The Most Probable End

Since the mass density has been determined to be very close to the critical density and the expansion rate of the universe has been found to be increasing, we now believe our universe to be very nearly flat with a small nonzero value for the cosmological constant.[11] While uncertainties remain as to whether our universe will ever suffer a Big Rip, the likelihood of a Big Crunch is all but eliminated. It will become more evident why this is important in the next chapter, as we consider why the Big Crunch was a favored model among cosmologists with an atheist agenda.

If we could stand and watch until the end of time, it would likely unfold like this. As the universe expands, gravitationally bound structures such as galaxy clusters would continue to shrink as gravity pulls member galaxies closer together. But the clusters themselves would be drifting ever farther apart at an accelerating rate. Eventually, each galaxy cluster would be isolated from the rest of the universe. Within the galaxy cluster, galaxies would continue to merge.

Stars would continue to form, consume their fuel, and die. When astronomers chart the rate of star formation in the past, they find a shocking trend. Star formation has been slowing down for the last 11 billion years. A careful tally shows that 95% of the stars that will ever form have formed already.[12] When the universe reaches about 100 trillion years of age, there will no longer be material available for any additional stars to form.[13] The smallest and dimmest stars burn slowly and could provide faint illumination for up to an additional 11 trillion years.[14] But then, the sky will grow completely dark as the last of the smallest stars fade away. By this point our local super-cluster will have condensed into one dark, giant galaxy. Rarely and briefly, a star would light up when stellar remnants merge or fall into a black hole. But even black holes emit radiation, and over great lengths of time, significant mass is lost. Finally, after an unimaginable length of time, all black holes will have evaporated, leaving only dispersed radiation. This represents the ultimate, inescapable end.

Chapter Three

The Fallacy of the Modern Assumption

When we run the cycles of cause and effect backwards, eventually you hit a brick wall at the Big Bang. By running the current motions of the galaxies in reverse to see what the universe looked like in earlier epochs, eventually they all meet at a common origin at the beginning of time. The laws of physics break down at this point, since density and temperature would have been beyond calculation within this singularity.

When you look forward in time, the universe as we know it, will end under all scenarios we can imagine. In light of the Big Bang, we know that the universe had a beginning. Remember that early in the twentieth century the steady state theory of the universe was prevalent in the scientific realm. This theory regarded the universe as eternal, having always existed and would never end. Remember also that the First and Second Laws of Thermodynamics have been well known for hundreds of years. How could scientists then favor a theory that would clearly violate these laws? At the time, it was believed that some exception would be found at the cosmological level to validate their view. Of course, it was not. Why then was this theory accepted? The reason comes from the "modern assumption" – *that everything is of natural cause and there is noth-*

ing supernatural that has had any effect on the natural world. This explains the resistance to the Big Bang theory when it was first proposed, since a universe with a beginning leaves room for a supernatural first cause.

Now there is use for the modern assumption in a small scale form, which is the scientific principle. It is advantageous to exclude supernatural effects in everyday evaluations of natural phenomena. This is valid because science is concerned with the study of the natural, not the supernatural. But because science does not deal with the supernatural does not mean that the supernatural does not exist. In questions of origin, the boundary between science and the supernatural become blurred. Excluding the supernatural cannot provide a complete evaluation of the topic.

Once the Big Bang model became well proven by the observation of galaxies and the discovery of the cosmic microwave background (CMB), those supporting the modern assumption had to come up with a new theory to explain the universe's origin. So the oscillating universe theory was born. Remember from general relativity, one possible end to the universe is the Big Crunch. Gravity could pull the whole universe back onto itself, recreating the singularity from which the universe began. This led to the notion that our universe began from the collapse of a previous universe, and that our universe will collapse to continue an endless cycle of destruction and rebirth. The process would repeat forever. This theory aimed to reinstate the idea of an eternal universe.

One problem for this theory was that the universe would need to have sufficient matter so that the density of the universe was greater than the critical density. As this theory came to dominate in cosmology, efforts to determine the density of the universe were underway. Though they could not find enough matter in the universe to exceed the critical density, it was assumed that the missing mass would eventually be found. Even with support for the existence of dark matter, there was not enough mass found to expect that the universe would ever collapse. Add to this the fact that an oscillating universe would, as the steady-state model did, violate the Second Law of Thermodynamics. This is essentially a universe-sized perpetual motion device. But despite these problems, this theory had broad support in cosmology simply because it was the only way to maintain belief in an eternal universe and the modern assumption of atheism.

In 1998, The Supernova Cosmology Project, which had begun expecting to measure the deceleration of the universe, instead discovered that the universe's expansion was actually accelerating.[1] Since the energy density has now been found to be very near the critical density, this acceleration is believed to be due to a non-zero value for the cosmological constant. The cosmological constant of Einstein's field equations was long thought to be simply zero. This is where dark energy comes in. We do not know exactly what it is, but we can see its effect on the universe's expansion. As further work has corroborated the fact that the universal expansion is accelerating, the possibility of a Big Crunch has diminished. So the oscillating universe model has lost support and is currently the least likely fate of the universe. Yet again, the modern assumption has steered cosmologists wrong.

I remember being taught the oscillating universe theory(cyclic model) when I was in school in the 1980's. It successfully detracted me from recognizing that the well-proven Big Bang theory actually supported a created universe. This is how support for one premise can be skewed with unproven counter explanations. There will always be ways to imagine alternate explanations for any evidence. The difference lies in those ideas that are supported by facts and those that are largely speculation. In this case, the oscillating universe theory was found to be an idea based on hope in a philosophical belief, not a sound theory based on facts. So the popular belief in creation remains a viable explanation of origin, at least in that the universe is known to have a beginning, albeit 13.8 billion years ago.

With the many failures of the modern assumption, you might expect it to lose credibility. Yet it still pervades every branch of science pertaining to origin. Those who depart from it are often ostracized within the scientific community. Belief in a creator is often considered to be based only on faith. But here we see that non-belief in a creator may require faith as well, since evidence for creation can be found within Big Bang theory and general relativity.

Chapter Four

The Fine-Tuned Universe

When we examine the laws of physics, we find that the laws of nature are entangled by several universal constants. Often these constants relate seemingly unconnected realms of science. Many scientists have marveled at the chance occurrence of certain quantities at just the right strengths to permit life in the universe.

These quantities consist of fundamental constants of nature and initial parameters of the Big Bang, whose observed values are not predicted by theory. Some of these are infinitesimally small and others are unimaginably large. We must measure the values of these parameters because theory cannot predict them. Quite possibly, these parameters could have taken on any value. But it is fortunate for us that they all have just the right values. Many of these parameters, if different, would preclude our existence as carbon-based organisms.

Some have proposed, however, that even if carbon-based life could not exist, perhaps another form of life might develop in its place. But the remarkable criticality of many of the physical constants of nature go beyond just biochemistry. It is doubtful that a random assortment of these constants could produce anything more than a universe consisting of dispersed hydrogen gas.

Universal Expansion and Density

From Big Bang theory, we recognize that the universe has been expanding since the beginning of time. When we study the equations that govern the universe's expansion, the outcome of the Big Bang was far from trivial. The initial conditions of the Big Bang critically affect the final result. The density of the early universe, the initial strength of the expansion, the strength of dark energy, and even the speed of light are all inputs into the equations that shaped the delicate balance in the universe.

The structure of the universe is largely dependent on the density of matter within it and the initial energy of the expansion. In chapter two, it was noted that the universe is very near the critical density. This is the density required for the expansion to continue forever but at an ever decreasing rate (excluding the effects of dark energy). The universe tends to fall away from the critical density over time. For the universe to be near the critical density now, after 13.8 billion years, it would have been extremely close to the critical density at the beginning. The initial density of the universe was within one part in 10^{60} of the critical density. Had the initial density been one part in 10^{57} greater, the universe would have re-collapsed in a Big Crunch within a few million years. If it were smaller by a lesser amount, the universe's matter would have rapidly dispersed before stars or galaxies could form.[1] This follows from a simple application of Einstein's field equations, and has been noted by several prominent cosmologists.

Now inflation theory may help to explain some aspects of this remarkable condition, but the mechanism for inflation has not yet been found. Inflation requires its own set of finely-tuned properties for it to function in way that produces the flat topology we require for a long lasting universe.

The discovery that the universal expansion is accelerating came as a great surprise to the world of physics. With it came the realization that the cosmological constant in Einstein's field equations has a nonzero positive value. The cosmological constant is acutely critical to the evolution of the universe. The fact that it has such a small value has permitted our existence. Cosmologists are quite puzzled by this fortunate state. A minutely larger value (by 10^{-119} change in the proportion of dark energy to matter energy) would have caused the expansion of the universe to be so rapid that the fragile density variations in the early universe would never have condensed into stars or galaxies.[2][3] Had the constant been negative (or less by 10^{-120}), the universe would have collapsed in a Big Crunch shortly after it began. This represents one of the most acutely fine-tuned parameters in physics.

Strong Nuclear Force and Deuterium

The strongest of the natural forces is the strong nuclear force. It is the force that binds groupings of quarks together in baryons (such as protons and neutrons). Residual effects of this force are responsible for the attraction between neutrons and protons at small distances, such as within the nucleus of an atom. At this scale its influence is greater than the electrostatic forces vying to push the nucleus apart. Its value determines the neutron to proton ratios that are stable in an atomic nucleus.

Within stars, new elements are synthesized and in the process energy is released. This energy powers the stars, making those like our sun shine bright and hot for billions of years. The first step in this process is the fusing of two protons to form deuterium(^{2}H).

$$^1H + {}^1H \rightarrow {}^2H + e^+ + \nu$$

Deuterium is a form of hydrogen that consists of a neutron and a proton in its nucleus rather than just a proton as in simple hydrogen. It fuses very slowly since this reaction does not release much energy, so it is difficult to achieve even in a stellar core. Once deuterium is formed, a rapid chain reaction converts this deuterium to Helium, and in the process it releases lots of energy.[4] So we see that deuterium is the key to making stable long-lived stars like the sun. But if the strong nuclear force were about 5% weaker, deuterium could not exist, since a single proton and neutron would not stick together. This critical step in the energy production of stars is necessary for forming helium, and helium is required for forming heavier elements. Without deuterium it is unlikely that stable long-lived stars could ever exist.[5] Not only do stars produce the energy to warm nearby planets, they also produce the elements by which planets are made.

On the other hand, if the strong nuclear force were only 2% stronger, a di-proton pair would form. It would then decay into deuterium via the weak interaction.[6] This would short circuit the slow formation of deuterium that moderates energy production in stars. Hydrogen would so easily fuse into helium that all of the universes hydrogen would have been consumed in the first few minutes after the Big Bang. No hydrogen would remain to power the stars, especially stable long-lived stars that allow for habitats such as Earth. Nor would there be hydrogen available for organic chemistry and water, which are essential for life. As it is, the strong nuclear force is just right for long stellar lifespans and limiting helium production in the first minutes after the Big Bang to about 24% of the initial matter.

Atoms and the Mass of the Proton

Normal matter in the universe is currently 74% hydrogen, 24% helium, and only 2% of other elements.[7] The universe's abundance of hydrogen allows for stars like the sun to be fueled for billions of years. Hydrogen is also important in organic chemistry and for life. Hydrogen makes up 10% of your weight, and is required in every organic molecule in your body.

Before the first atoms formed shortly after the Big Bang, sub-atomic particles formed as the energy density of the universe dropped in the early stages of expansion. Electrons, neutrons, protons, neutrinos, and all of their antiparticles formed in great abundance. Physicists can easily calculate many conditions from this time period, especially the abundances of each particle produced as the energy density dropped at a predictable rate.

Early on the creation and destruction of these particles occurred in equilibrium due to the high temperatures and pressures. As the universe expanded, it cooled, with certain reactions stopping when it became too cool for them to proceed. The resulting cosmic neutron/proton ratio can be found by using the Boltzmann factor, $e^{(m_n - m_p)c^2/kT}$.[8] In this formula, T is the temperature that it became too cool for the conversions between neutrons and protons to be sustained. This factor relates the proportion of neutrons and protons created by considering their mass energies, resulting in about one neutron for every seven protons produced during the Big Bang.

However this ratio is highly dependent on the masses of the particles. If the mass of the neutron were 1% larger or the proton were 1% smaller, then about 50% neutrons would have been created during the Big Bang. This gives an equal number of each. For this case, or any case with more neutrons than protons, all protons would be paired with neutrons.[9] And all hydrogen would be in the form of deuterium, which would rapidly fuse into helium in the first minutes after the Big Bang. The universe would have began as all helium, leaving no hydrogen to power stable long-lived stars, or for hydrogen containing molecules like water, necessary for life.

The situation is no better for the opposite condition, having a smaller neutron or a larger proton. If the neutron were lighter or the proton more massive by just 0.2%, then free protons would decay into neutrons, in which case there would be no atoms at all.[10] So we see that the ratio of the masses of the proton and the neutron are in perfect balance to allow the great diversity of elements we observe and the abundance of hydrogen in the universe.

Fine-Tuned for Carbon-Based Life

Though it has been argued that some non-carbon-based life may exist somewhere in the cosmos, we must perceive that this is rather unlikely. No one can deny the incredible properties of carbon. Carbon is the only element which can easily form long chains, which results in a multitude of organic molecules when combined with other elements. Carbon is so well suited for making molecules, that there are more molecules that contain carbon than those that do not. The chaining nature of carbon allows for long molecules such as DNA, proteins, hydrocarbon fuels like gasoline, plastic compounds, and many many biological substances. We cannot say for certain that no other chemical foundation for life is possible, but we can see that carbon is, hands down, the best element for the job. So much better that many would believe that life of any kind would be impossible without it.

Carbon, however, does not act alone. Hydrogen and oxygen are also very important in organic chemistry. Water (composed of one oxygen and two hydrogen atoms) is a universal solvent that is exceptional at facilitating the creation of organic, carbon-containing substances. Water is so important to life that nearly all searches for extraterrestrial life focus on the search for planets where liquid water could be present. It is doubtful that there could be any form of life that does not require liquid water.

The five most abundant elements in the universe by the number of atoms are hydrogen, helium, oxygen, carbon, and nitrogen. Excluding helium, which is inert and does not react chemically, these most common elements make up over 96% of your body weight.[11] The chemical composition of all life is made up of the most abundant elements in the universe. This is not because life is evolved to use the most common elements; carbon-based organics represent the only chemistry known to work for life, and are by far the best suited for the complex chemistry necessary for life. It is remarkable that the best elements for making life also are the most common in the universe.[12]

The Triple Alpha Process

The famous astronomer Fred Hoyle studied the nuclear physics behind element formation in stars. He was puzzled by the fortunate abundance of the elements. Since the Big Bang produced only hydrogen and helium, all other elements must be formed in stars, the only natural powerhouses capable of driving the fusion of elements. Stars produce energy by converting hydrogen into helium over long periods of time. But when hydrogen is exhausted in the core of a star, there is an energy barrier that inhibits further fusion. Yet carbon is abundant in the universe. The only way that carbon can originate in stars is through the combination of three helium-4 nuclei via the triple alpha process. This is a problem, since simultaneous collisions are nearly impossible. However, if an intermediate step existed, then it would be possible. If the star is massive enough, beryllium-8 can be formed from two helium-4 nuclei. But this isotope of beryllium is highly unstable and so short lived that there is little chance of this particle fusing with another helium-4 before it breaks apart.

Hoyle reasoned that the only way beryllium-8 and helium-4 could fuse in time is if there was an excited state of the carbon-12 nucleus that was at the same energy level as the two constituent particles. He calculated the energy level required and asked researchers to check for it. This excited state was found experimentally shortly after, and with the expected energy of 7.654 MeV.[13] Atomic nuclei have very few and very specific excited states. The fact that these states happen to align favorably in support of carbon fusion is baffling. Hoyle, who before this discovery was a stanch atheist, recanted, citing this unlikely alignment as evidence for intelligent design in the laws of physics. Without this finely-tuned condition, carbon would be almost nonexistent in the universe. Elements beyond carbon would not be formed in stars either. Without these elements, life would be impossible.

The Balance between Carbon and Oxygen

Neither carbon nor oxygen could be missing from the universe, if life was ever to exist. Both are required in the chemistry of life. We have already seen how the triple alpha process is finely tuned to allow for carbon production in stars. Once carbon is made, the formation of other elements is possible by the further fusion of the elements with helium. Carbon can fuse with helium to form oxygen, oxygen can fuse with helium to form neon, and so on.

The physics behind nuclear reactions for each step in element production are known to the degree that models can be made that make accurate predictions for the rates of these reactions and the resulting abundances of the elements. When various sizes of stars are modeled, the influence of two fundamental constants have been evaluated. The strength of the strong force, which binds neutrons and protons together in the atomic nucleus, and the strength of the coulomb force, which acts between electric charges, both critically affect the rates of these reactions. When the calculations are done to evaluate what kind of rates would occur if the constants that control these forces were different, some astonishing results are obtained.

The coulomb force, could not be greater by more than 4% or the carbon burning reactions would be too efficient, converting nearly all carbon to oxygen as fast as it is produced. On the other hand, if this constant was lower by 4%, carbon burning would be highly suppressed, producing hardly any oxygen.[14]

Similarly, the strong force could not be less by more than 0.5%, without transforming nearly all carbon into oxygen. On the other hand, if the strong force were greater by 0.5%, then it is oxygen that would be scarce in the universe.[15]

As it is, these forces are optimized to just the right values to allow for generous abundances of both carbon and oxygen in the universe. There is a delicate balance required for stars to produce oxygen without consuming all of the carbon. A small deviation in these parameters from their actual values would have resulted in the universe lacking one of these critical elements. Both constants are just what is required to allow for life.

The following table shows several of the universe's critical parameters and the maximum variation the actual value that could be tolerated without preventing life in the universe.

Examples of Fine-Tuning		
Fine-Tuned Parameters	Max Variance for Life	
	(+)	(−)
δ Universe Initial Density	$1/10^{57}$	$1/10^{60}$
Λ Cosmological Constant	$1/10^{119}$	$1/10^{120}$
$\dfrac{m_n}{m_p}$ Neutron to Proton Mass Ratio	$1/100$	$1/500$
g_s Strong Nuclear Force (For Fusion of Helium)	$1/50$	$1/20$
(For Fusion of Oxygen)	$1/200$	$1/200$
k_e Coulomb Force	$1/25$	$1/25$
^{12m}C Carbon-12 Excited State	$1/100$	$1/100$

Designed for Life

In the sci-fi classic movie, *The Matrix*, hackers discover that something is wrong with their world; things seem to be rigged. In their search for an explanation they discover they are living in an artificial reality. A world created to imprison them.

As science has embraced the modern assumption, researchers expect that natural explanations are all that is required to explain reality. But like the hackers of the Matrix, many are bewildered by the fact that so many properties of our universe seem to be rigged. Something appears wrong from a naturalistic perspective. This phenomena is known as "fine-tuning", and examples of it are too numerous to list them all here. Scientific journal articles and reports are littered with examples of this, and usually regard it as a "problem" needing to be solved, since it is a problem for the naturalistic point of view.

When we examine the Earth, we can't help but marvel at its unique life-giving properties. This is even more remarkable with study of other star systems and planets. The many requirements placed upon a planet, for it to support life for billions of years, makes planets like the Earth seem to be exceedingly rare. But this rare world might seem as no surprise, given the multitude of planets in the universe. When we consider fine-tuning, however, the remarkable state of nature is universal. The whole universe shares the same laws of physics; the same laws that are fine-tuned for complexity and for life. If not for these, the whole universe would be devoid of life.

The universe and all the material within it are governed by the laws of physics and the equations they prescribe. Many physicists have pondered, "Why are the laws of physics the way they are?" In light of the previous survey of some of the physical constants which steer the equations, one is drawn to an inescapable conclusion. The laws of physics are perfectly crafted to allow for complexity, life, and even intelligence. It could be said that we are extraordinarily lucky that the laws of physics are just right for us. Or we just evolved to fit the conditions of the universe. But this does not explain how the universe is just right for atoms or stars or galaxies. Life could not exist in any form without these. Neither physicists nor philosophers like to rely on luck, so we must conclude that there is a reason for this fortunate state of being.

The most obvious reason for fine-tuning is that the universe is the work of an intelligent creator, whose purpose was the building of a richly complex universe with stars, galaxies, and life. The God-designed balance of nature has been the dogma of theologians for millennia. Now science has quantified this perfect balance with great precision. This has led some scientists, such as Hoyle, to come to a belief that the universe was intelligently designed. This represents one of the most foundational pieces of evidence for creation by God that can be found through science.

Chapter Five
Multiverse Theories

The notion of parallel universes has been the basis of some captivating story lines in many science fiction narratives. But in recent time this idea has received attention from some cosmologists as well. This stems from the astounding degree of fine-tuning that is required for life to exist. By making use of parallel universes, the naturalist has sought a way to explain the fine-tuning of the universe without invoking divine influences. Just a few extra universes could not explain our existence, but an infinite number are required. This represents the concept of the multiverse, an ensemble of an infinite number of universes, of which ours is one.

This model has many variants, but all propose that there exists infinitely many universes, each with some variation in properties. These multiverse theories aim to answer the question, "Why is the universe just right for life and for intelligent beings like me?" According to multiverse theory, the answer is this: "Our universe is ideally suited for us because we could only exist in such a universe. In any universe unfit for life, there will be no one to ask the question." This implies that the vast majority of universes are simple and uninteresting and devoid of life. But with an infinite number of universes to choose from, there will be a minute fraction that have great complexity, atoms, molecules, stars, and intelligent life.

Max Tegmark has categorized multiverse theories into 4 types or "levels" and described some of the consequences of each.[1] String theorists have also come up with their own multiverse theories. All of these models lead to some remarkable outcomes, if indeed any of them represent an accurate account of reality. For instance, the existence of any level of multiverse requires that there be infinite copies of you throughout the ensemble. Not only exact copies, but also variants with slightly different characteristics or life histories. Any conceivable occurrence, such as fictional stories will also be realized somewhere in the multiverse. The consequences of having a truly infinite number of universes can be somewhat disheartening, even to an avid sci-fi fan.

The Level 1 Multiverse: An Infinite Universe

Now we will take a brief look into the first of the multiverse levels and evaluate its merits. The level 1 model proposes that our universe is infinite in size, with separate Hubble volumes making up the individual "universes". When we observe the heavens, we come upon a limit to the distance we can see. Not because our telescopes lack the power or resolution, but because light from more distant parts has not had time to reach us in the life of the universe. The part of the universe we can see, the observable universe, is known as our Hubble volume. The more distant regions are out of causal contact, since light cannot interact with us, and no interaction is faster than light. An observer in another of these regions would also be limited by how far they could see and would have their own Hubble volume.

In this multiverse concept, the separate Hubble volumes make up the individual universes. The individual Hubble volumes are arbitrary and result from the perspective of the observer. Someone just beyond our horizon would have a Hubble volume that overlaps with ours, so the larger universe is still seamless. As time progresses, each Hubble volume will grow in size by one light-year per year as more light makes its journey to us. But due to the accelerating expansion of the universe, our Hubble volume will grow in size but shrink in content, as distant galaxies are pushed out of view. This is true of our Hubble volume, whether or not we live in a finite universe.

It is the notion of an infinite universe that can generate the strange consequences of a multiverse without requiring that completely separate universes exist. Though the laws of physics would be the same for all Hubble volumes and all would have a common origin at the Big Bang, the initial condition of each local region could be different.[2] Density and matter distribution could be different. Some have speculated that some other parameters and constants could vary through space and time. With an infinite number of Hubble volumes, there would exist others identical in every way to ours, as well as those that are similar but slightly different. Indeed, every possible history permissible to the laws of physics would be played out somewhere in the infinite universe.

The level 1 multiverse is popular among some cosmologists because it can fit within current observation. To be certain, we cannot yet say if the universe is finite or infinite. Measurements of density fluctuations in the cosmic microwave background indicate that the curvature of the universe is less than our current ability to measure it.[3] This suggests that the universe is either totally flat or the degree of curvature is too small to be measured. Any degree of positive curvature would indicate that the universe is not infinite but would wrap back around on itself as does the surface of a sphere. Though flat geometry would allow for infinite space, there are four dimensional shapes that are geometrically flat but still finite in volume. An example of this is the 3-torus, a four dimensional version of a doughnut. The 3-torus has received a lot of attention as a possible shape of the universe. But if space is flat and infinite, then the frightening consequences of the infinite universe model may apply. However, if the universe has any degree of curvature or its topology resembles a 3-torus, then you are unique in the cosmos.

The Level 2 Multiverse: The Inflating Background

There are many variants of the level 2 multiverse that have been proposed. These models postulate that there exists an infinite number of universes, each with its own set of physical constants. Though the laws of physics may be the same in all universes, the basic parameters like the speed of light, the gravitational constant, the charge of an electron, etc. could vary between realms.

As with the level 1 multiverse, these models are very difficult (if not impossible) to test for. There is no evidence for their existence. However, there are theoretical models which provide grounds for consideration of the level 2 multiverse. One such model that Tegmark sights is called "chaotic inflation".[4]

Inflation theory is an extension of the Big Bang model. It proposes that shortly after the Big Bang, the universe went through a period of extremely rapid expansion. This expansion may have lasted only a fraction of a second, but in that time the size of the universe increased by a factor of at least 10^{26}.[5] By contrast the slow expansion that followed has only yielded an increase of about 1100 times since the CMB was emitted.[6] Inflation explains why the universe is so flat since it would have increased the size so much that any curvature would be undetectable on small scales. It also explains why the cosmic microwave background is so uniform in temperature. Before inflation, nearby regions would come into thermal equilibrium. Then inflation would spread this isothermal region far beyond the edge of the observable universe. Since the early universe was so uniform, it is remarkable that areas existed dense enough to allow for galaxy formation. But quantum fluctuations during inflation would have been stretched to macroscopic size by the time inflation ended. This provided the slight density variation needed to spur galaxy formation in the early epochs of the universe. With all of its successes, it is no wonder that inflation theory is widely embraced.

Now chaotic inflation is an entirely different model. It proposes a sort of perpetual inflation that once started, will continue forever. But occasionally, a region stops inflating and becomes a universe of its own. This creates a bubble within the inflating background. This bubble could be infinite in size and there may be an infinite number of them. These bubbles make up the level 2 multiverse.[7] This theory challenges the Big Bang model, proposing that the universe is much older than 13.8 billion years. The Hubble age measured by cosmologists is then just the time since our bubble stopped inflating. This theory also proposes that the conditions of each bubble would be different, yielding a new set of physical constants for each bubble universe. Tegmark sights the many examples of our universe's fine-tuning as evidence for a selection effect within an infinite ensemble of universes.[8]

Similar to chaotic inflation is eternal inflation. The mechanism behind it differs, but the basic outcome is the same, inflation that lasts forever, creating an infinite number of universes in its wake. Both of these theories require very specific characteristics of the inflaton field. This field represents nature at a more fundamental level, since it is the source for variation of the parameters of physics within the universes it creates. Both of these theories are very speculative and are distinct from non-eternal inflationary theories.

The type of field required for inflation is called a scalar field; this is a field that has a numeric value for its strength but does not have any direction associated with it. The scalar field associated with inflation is hypothesized but has not yet been observed in nature. The characteristics of this scalar field must satisfy very special conditions in order to produce inflation and very peculiar requirements for it to maintain itself for eternity. For eternal inflation, it requires that the field have a minimum energy value when the field strength is at some critical level above zero. This means that the field is more energetic when it is not present than when it is present at this critical value. Inflation theory also requires that this field must decay slowly, giving inflation time to act before the field dissipates.[9] The energy density in space must remain constant, even as space expands, for inflation to occur either briefly or eternally. Eternal inflation theories depend on these very strange and special requirements of the inflaton field. These requirements indicate another degree of fine-tuning that would be necessary even at the multiverse level.

The Level 3 and 4 Multiverse: Everything Exists

The level 3 multiverse postulates that all quantum events that could occur do occur. Each quantum event splits the universe into identical copies that differ only by the outcome of a single quantum event.[10] By this notion, every possible history represents an alternate universe. No particular one is any more real than the others, except to the individuals within it.

The level 4 multiverse consists of varied laws of physics. This says that anything that can be described mathematically actually does exist.[11] This provides for an infinite number of universes with differing equations of physics as well as the constants. This is the mathematician's dream, as any set of equations he can dream up will describe a real universe somewhere in the multiverse. Perhaps the act of dreaming up the equations actually creates that universe?

Brane World Hypothesis

Proponents of string theory and M-theory postulate that reality is made up of many more dimensions than the three spacial and one time dimensions we are familiar with. In this context it is possible to imagine alternate universes existing on planes of reality that we cannot access. These alternate universes are referred to as branes, with the whole collection of branes being the brane world. Notable proponents of the brane world hypothesis are Brian Greene, Stephen Hawking, and Leonard Mlodinow. Greene has stated that the brane world arises out of the mathematics of string theory. He describes these alternate realities as being like giant slices of bread, each universe only inches apart but inaccessible none the less.[12] The brane world has all the characteristics of the other higher level multiverse theories, containing infinitely many universes with varied properties.

The brane world theory has recently lost some traction with a recent failure of string theory. The most basic string theory models predict supersymmetric particles, that are heavy versions mirroring all normal particles. Scientists at CERN, the world's largest particle accelerator, have recently determined that these supersymmetric particles simply do not exist.[13] [14] This puts string theory, M theory, and the brane world on shaky grounds. Not only do these ideas lack any physical evidence, but now one of string theory's central predictions is shown to be false. Without string theory, the even more speculative notion of the brane world multiverse has lost merit.

This major failure of string theory and the brane world hypothesis highlights the truly speculative nature of theoretical physics. This realm of physics conducts an important roll in probing into the unknown. But we cannot count on these theories as a foundation of any world view, since each theory's chance of success is slim. It turns out that the other multiverse scenarios do not fare any better, all of them relying on very speculative theories that still lack supporting evidence.

Flaws in the Multiverse Position

Probably the most popular multiverse scenario among astrophysicists is that of the infinite universe. Our universe is known to be very large (at least 100 billion light-years across), and no maximum size has yet been determined. But to say it is infinite because we have not yet been able to measure any curvature, and hence its size, is a bit premature. There is no mechanism for something finite to become infinite. So if the universe is infinite now, it must have begun with infinite size.[15] Inflation explains the flatness of the universe due to its rapid growth to the point that any curvature would be vanishingly small.[16] So we would expect from standard inflation theory that if the universe is finite, it would be so large and so nearly flat, that it would be observationally indistinguishable from being infinite. This leaves us with the realization that we may never be able to determine the true extent of the universe.

Even if our universe is infinite in size, it does not solve the fine-tuning dilemma. The fundamental constants of nature would need to vary throughout the cosmos in order to offer any natural explanation for the observed level of fine-tuning. Quite the opposite is true, though many attempts have been made, no convincing evidence for variation in time or space of the fundamental parameters of nature have ever been found.[17] [18] The fine structure constant is a relation between the electric charge, the speed of light, and Planck's constant. By comparing the spectra of very distant quasars, physicists have found that any variation in the fine structure constant, across the visible universe, would have to be less than one thousandth of one percent. Some evidence of variation has been claimed by an opposing study, but it is an order of magnitude smaller than the error margin of the measurements.[19] This is basically a testament to the constancy of the constants, within the limits of our ability to measure them. Another groundbreaking discovery came from analysis of alcohol molecules in distant galaxies filtering light from more distant sources. This study showed that the ratio of proton to electron mass has not changed in the seven billion years since this light interacted with the foreground galaxy. The level of accuracy of this result is to one part in ten million, the best measurement yet taken in the search for variation of the constants.[20] That is a profound level of accuracy. Now we know that particle masses, the speed of light, the electric charge, and Planck's constant are all found not to vary in time or space as far as we can measure. Infinite or not, the universe is still fine-tuned for life.

When we examine multiverse levels 2 through 4, we must rec-ognize the truly speculative nature of these topics. Proponents of the level 2 multiverse try to ascribe the successes of standard inflation theory to eternally lasting inflation. But inflation is a very specula-tive topic, and there are as many different theorized forms as there are physicists. To equate the fraction of a second period of inflation that observations support to an infinite and eternal inflating back-ground is quite a stretch. Some physicists have proposed strange requirements to the field that causes inflation so it could perpetuate forever. However, this would violate the First and Second Laws of Thermodynamics (conservation of energy and entropy must always increase). Even if other universes do exist, we would expect these foundational laws to be regarded, whether or not the constants or equations differed. (They are based on simple mathematics.) Not to mention that this very strange requirement is counter to every other kind of field known, in that energy density always decreases when it is expanding. Inflation in our universe likely ended when the energy driving this accelerated expansion had dissipated to the point that it could no longer be sustained. If the inflaton field maintained a con-stant energy density, then some other field or energy source must have been decaying to sustain it. But that source field would eventu-ally wane and inflation would end.

It has been proposed that conservation of energy may not be violated by inflation if negative energy of the gravitational field exactly balances the increase in particle energy.[21] But this still neglects the Second Law of Thermodynamics that forbids a decrease in entropy. By creating a larger and larger negative energy field and at the same time a matching amount of positive energy in the form of particles, a vast potential for work has been created. In essence a decrease in entropy. Unless this is driven by an initial store of energy at lower entropy, which runs out in a finite amount of time, then this scenario would violate the Second Law of Thermodynamics. Over the last century, cosmological model after model have proposed second law violations, and each time these theories were eventually found to be false. There are no perpetual motion machines, not on Earth, nor anywhere else in the cosmos. Eternal inflation is another perpetual motion device that will be rejected when enough data becomes available.

Alan Guth, an author of the original theory of inflation and proponent of eternal inflation, has stated another limitation of his own theory. While he supports eternal inflation as being eternal into the future, he points out that it cannot be eternal into the past.[22] In other words, eternal inflation must have a beginning. So, even if eternal inflation pushes back the beginning of time so that it precedes the Big Bang, the multiverse still requires a creation event that is beyond the bounds of science. This multiverse still depends on the fine-tuning of the properties of the inflaton field, in addition to the existence of quantum laws of nature and leaves the origin of the initial space-time unexplained. A fine-tuned creation mechanism at some point in the past is still required for this form of multiverse. So even if one wishes to believe that future discovery will find the inflaton field to match the requirements of eternal inflation, this multiverse scenario will still require an intelligent creator to give it the precise properties required to produce universes. We would also depend on this creator to get it all started in the first place.

The universe is by definition a closed system. This makes direct testing for multiverse levels 2-4 impossible. The multiverse framework is often used to provide a purely naturalistic explanation for the universe's fine-tuning. But observational support for such ideas are lacking. However, fine-tuning and the finite age of the universe are well established and provide strong support for the design argument. The numerous multiverse theories are contrived to explain away the fine-tuned nature of the universe and are conveniently untestable.

I have listened to many lectures by prominent theorists such as Guth, Greene, Carol, and others. It is remarkable to me how often the subject of God comes up in these discussions. They also admit the shaky grounds of the multiverse notion, but usually state that when all other explanations have been exhausted, this is all that is left. The far-fetched notion of a multiverse is considered only because these theorists have dismissed any consideration for the existence of God. The multiverse is the only explanation that atheism can produce to counter the established evidence for fine-tuning.

From the multiverse perspective, fine-tuning is the result of a selection effect. Only universes that have the appearance of fine-tuning will be capable of hosting intelligent observers who can ask, "Why is the universe just right for life?" But consider a man receiving the death penalty by a firing squad. A hundred riflemen all shoot at the same time, and to everyone's surprise, they all miss. The prisoner is released and ponders, "Why did they all miss; did someone intentionally act to save my life?" Clearly this man would not be around to ponder the question if all one hundred had not missed. But that does not mean that chance alone is a satisfactory explanation. There must be a reason for his survival. Certainly, someone tampered with the riffles or the munitions, or the executioners were bribed. Whatever the reason, good luck does not explain it.

The same goes for the fine-tuning of the universe. Multiverse theories propose we could only exist if conditions are right for our existence. Our just-right universe is a condition too perfect to leave unanswered. This fortunate situation for life could not be purely by chance, but must be for a reason.

While the prospect of alternate universes is tantalizing, there is little support scientifically. This is not to say that there could not be other created universes or even a created multiverse. This would be quite possible, if you accept our universe as created. But infinitely many, accidentally occurring universes represent a theoretical muse that distracts from credible elements of cosmology. Science deals with phenomena that is measurable, testable, and predictable. So our universe represents a boundary to science, outside of which is the realm of religion or philosophy.

The simple and most obvious explanation for fine-tuning is that the universe is intelligently designed. This is exactly what you would expect to find for a created universe. Long before we could calculate the stability of a stellar core or the charge of an electron, the perfect balance of the created order was predicted from the theistic viewpoint. Now this balance is measured with precision by science and is in perfect agreement with creation theology.

Chapter 6
Quantum Mechanics

We have reviewed many of the laws of nature that govern the universe. Does the orderly and predictable nature of the universe seem too mechanical to have been created by a loving and caring God? Do these laws control the flow of events so rigidly that there is no room for action by God? Is there any place left for free will when you consider the complex cocktail of chemical reactions within the brain that are also bound by these same natural laws?

Determinism

Given our current understanding of the natural law that allows us to calculate spacecraft trajectories, perform precision eye surgeries, and harvest power from nuclear reactions, you might think that science could one day foretell any future event if the initial conditions are precisely known. By this thought, all matter in the universe has been following strict natural laws since the beginning. This chain of cause and effect strings unimpeded from the present all the way back to the Big Bang. This leads to the outmoded principle of determinism. The principle of determinism postulates that if you knew all physical laws and the exact state of things at any particular time, then you must be able to predict all future events by applying the physical laws to the data of the known time frame. Any gap in your ability to calculate the future is due to having incomplete knowledge of the present situation. This leaves no room for chance or even choice, since all events, even our thoughts, are just the natural evolution of the physical laws. If you could roll time back and start it all over at the Big Bang with the exact same starting conditions, the universe would repeat itself exactly the same all over again.

Now this would seem like a very impersonal creator, to create the universe and then stand back and just watch from a distance. This is essentially deism. The deist would say that there is a god but he currently has no influence in the universe and quite possibly has no interest in us. If the universe were totally deterministic, one might ask, "Is there any room for action by God?" or by extension, any need for God at all? Determinism supported both deism and atheism for a time, but then came quantum mechanics.

The Uncertainty Principle

Quantum mechanics is rooted in the fact that at the small scale energy is quantized. That is, it comes in discrete quantities, that cannot be divided infinitely. With the discovery of the atom, we found that all matter was made up of particles. Atoms themselves are divided into smaller particles of neutrons, protons, and electrons, and the first two can also be broken down further into quarks. It is at this level that we find particles that are truly fundamental and are not made of anything else besides energy. So electrons, quarks, neutrinos, photons, and a handful of others make up the list of fundamental particles. Finally, we find that even energy comes in bundles, and only certain quantities are allowed, depending on the medium containing it.

The world of the extremely small gets even stranger due to the Heisenberg Uncertainty Principle. This states that the values of certain properties of a particle cannot be known to just any degree of accuracy, but that the certainty of any measurement is limited by the relation:

$$\Delta p_x \; \Delta x > h/2\pi$$

(Δp_x is the uncertainty of the momentum of the particle, Δx is the uncertainty of its position, and **h** is plank's constant.)[1] This formula states that the uncertainty of the position multiplied by the uncertainty of the momentum must be greater than Plank's constant divided by 2π. The standard deviations are the uncertainties in this formula. This shows that any measurement has uncertainty associated with it that must be greater than a minimum allowable value. All this is to say, that there is a degree of randomness in the world. When doing experiments on subatomic particles, statistics must be invoked to handle these random tendencies. To measure the momentum of a particle, knowledge of its position is sacrificed. Conversely, the more you constrain the particle's position, the more uncertain the momentum will be. This is not a limit in our understanding or ability to take measurements accurately but a fundamental property of nature.[2]

Another way to describe this process is that of a wave function. The quantum wave function is the set of all possible states a particle could be in. These states are theorized to exist simultaneously for a time, then sometime between its last interaction and observation of the particle, the wave function collapses into one of the possible states. Some believe it is the observation that causes the collapse, while other physicists propose that the wave function collapses within a short time after the interaction, but we do not observe the collapse until we have taken a measurement. In either case, the collapse of the wave function yields an incalculable result, whose outcome cannot be predicted by science.

The implications at the particle level are huge. This means that two particles passing near each other can either pass by only interacting by their electric fields, they can collide and recoil onto totally different trajectories, or they can be destroyed and form new particles. You could rerun this scenario many times, and you will get different results each time. The random effects at the microscopic level often just average out for larger systems, but many times these effects accumulate to have a measurable effect.

As a result of Heisenberg's Uncertainty Principle, determinism is dead. If you turned back time, started the universe out with identical initial conditions, and then allowed time to run forward again, the universe would not repeat itself. Things would happen differently, since random effects control the interactions of all particles. It is therefore impossible to predict future events with absolute certainty even if you knew all the laws of physics and had complete knowledge of the system in question.

Such quantum effects do not need vast scales of time to become amplified enough to have observable macroscopic consequences. A recent study from the University of California at Davis probed the relationship between classical uncertainty and quantum uncertainty. They determined that all macroscopic (classical) probabilities result purely from quantum influences. They go on to show that in a coin toss, that quantum effects will swamp any chance of predictability.[3] This is especially true in any case where a human must act, since quantum effects influence our thoughts, our senses, the strength of a given electrochemical signal, and our reflexes. Any animal nervous system can be seen as an amplifier of quantum noise. Fluids are also an amplifier, which greatly affect the weather, solar activity, plate tectonics, and many other common processes. This shows us how influential these random quantum effects are on everyday occurrence.

We now must face the limits of science. There is no theory known or unknown that can predict certain quantities with greater precision than what is allowed by the uncertainty principle. These small random effects at the quantum level might seem trivial. But as we can see, random quantum effects can quickly dominate any natural process. Everything around us is made of particles. The actions of these particles are governed in part by these quantum effects. Quantum fluctuations gave rise to small density variations in the first moments after the Big Bang. These density variations were amplified during inflation to become the seeds of galaxy formation in the early universe. So the existence of our own galaxy is the result of the random preference of tiny particles to gather more densely into our region of the universe early in that first second of history.

Why are the laws of nature so strange at the microscopic level? Could there be a hidden mechanism controlling these seemingly random interactions?

A Mechanism for the Supernatural

If the Creator wished that the laws of nature be followed strictly at all times, not wishing to make laws and then violate them by divine action, then the uncertainty principle would leave an open door. By acting at the quantum level, the Creator could always have complete control without violating any natural laws. By this notion, the acts of God would be scientifically indistinguishable from the forces of nature, since divine action would be masked by the uncertainty principle.

This could be called the "fine-tuning of events". The laws of nature are the commands of God. And they are so crafted to fulfill his desire. Since He is perfect, He has chosen not to break his own law, but acts within it at the quantum level, maintaining complete control over all creation.

Since (random) quantum effects play a huge role in the final outcome at all levels, universal, galactic, and even at the planetary level, any desired outcome cannot be assumed by conditions set at an earlier time. So it becomes clear that the principles of quantum mechanics not only allow for intervention by God, but require it. For God would need to be continuously in control of these quantum effects to ensure any final result. Otherwise, random effects would quickly nullify any effort made in the past to affect the future. This portrays a picture of God expected by theologians, that God is in control of everything at all times.

The randomness of the quantum realm was a huge surprise to scientists and physicists alike. Many did not want to accept it, like Einstein, but as the evidence mounted, it was obvious that nature behaved very strangely at the microscopic level. This strange behavior was not a natural expectation for the physical laws. But if the universe was created by God and if He wanted to have a built in control mechanism, this would be the perfect solution. Masking the supernatural in this way allows nature to proceed coherently, yet still allowing for divine control. This allows us to discover the laws of nature and use them to our advantage, though God may take control of any final outcome.

In the Bible, God is described as acting many times throughout history. Most often, these acts are orchestrated through the control of natural phenomenon. Jesus calms the sea, Jonah is swallowed by a great fish, Noah escapes heavy rains and flooding, Daniel is saved from lions choosing not to eat him. Control of weather is control of the gases that make up our atmosphere. Thought is a quantum event, with microscopic chemical and electrical interactions. An animal's actions could be controlled if their thoughts were controlled at the quantum level. Healing, visions, celestial events, and every biblical miracle could be orchestrated through quantum actions taken by God at the appropriate time.

Does this seem too limiting for an all-powerful God? Power over the quantum world is power over everything. Anything can be accomplished with this kind of power, especially if the controller can act throughout time while having complete knowledge of the future. It may not be the only mechanism that God uses to act. But this seems to be the most probable method of choice for an all-powerful God to carry out his will and still maintain the natural order.

Psalms 139:13 states: "For you formed my inward parts; You covered me in my mother's womb." (NKJV). People of biblical times knew that a woman could become pregnant only by sexual intercourse and that the baby would take nine months to grow inside her. They knew that the child contained the "seed" of each parent and would be in their likeness. Yet knowing that natural reasons explain a person's traits and conception, the psalmist states here that it is God who forms each individual. This is because it is God who designed the natural law, and He continuously controls the forces of nature at the most fundamental level. This harmony between natural law and God's control was recognized even in biblical times. Only now we observe the scientific mechanism by which this takes place.

Maxwell's Demon

Recall the Second Law of Thermodynamics which requires that energy always flows from high concentration to a lower concentration. That is, energy tends to disperse (entropy always increases). It cannot go the other way unless another low entropy energy source is sacrificed. If you could violate this law, you could create a perpetual motion device that could supply unlimited amounts of energy.

James Maxwell in 1871 proposed a hypothetical experiment. Making use of the fact that temperature is a measure of average molecular motion and that individual molecular velocities vary throughout a gas, he devised a scheme for sorting these molecules. Starting with two sealed chambers (A and B) that were connected by a small trap door, single molecules could be allowed to pass through selectively by opening or closing the door.[4] He wondered if a small being was mischievously operating the trap door, such that when a fast molecule in chamber A was approaching the door, he would open the door and allow it pass through to chamber B. Similarly, if a slow-moving molecule in chamber B approached the door, he would open the door and let it pass into A. In so doing, chamber A would get cooler and chamber B would get warmer, without providing any energy to the system. This heat difference could be used to operate a simple engine without an external energy source. (See figure below.)

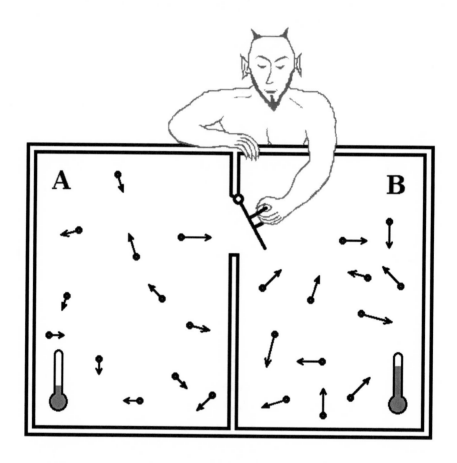

Now this would clearly violate the second law, but it presented quite a puzzle as to why. Physicists of Maxwell's day did not have an apparatus that could be used to monitor and select individual molecules, so direct tests could not be done. Many ideas were proposed, but it would be nearly a hundred years before a complete explanation would be found. This is undoubtedly why Maxwell's unnamed being would become known as Maxwell's demon.

It was quickly realized that the complete system had to include the demon, since it did act on the system. It was recognized that for the demon to do his job, he would have to do several thermodynamically significant tasks. He would take measurements of the speed of each molecule as it approached. He would need to remember this information for some arbitrary amount of time. He would also need to choose whether or not to open the door. He would then need to act on his decision. Finally, in order to return the demon to his original state, he would need to forget the measurement after it was used and no longer needed.

Over time, all of these tasks except one were found to be thermodynamically reversible (possible to be done without requiring the consumption of useful energy). Surprisingly, it is the erasure of information that is responsible for the rise in entropy (the conversion of useful energy to a less useful state).[5] If we replace Maxwell's demon with a super-efficient computer capable of carrying out the tasks, it will acquire information, then use the information to generate useful energy. But the experiment could not continue indefinitely, unless the computer's memory is periodically erased to make room for new measurement data. In so doing, energy is spent, at least as much as is gained by the demon's engine, in accordance with the second law.

The paradox revealed by Maxwell results in a profound conclusion. Equivalence to energy is not only a property of mass. Information is also a form of energy. Knowledge really is power. The theoretical equivalence of energy and information is well demonstrated by this application of the second law.

Researchers exploring this phenomena have produced a simple device akin to Maxwell's demon.[6] By using a series of electric fields and a nano-sized elongated bead which can rotate, they created a system were rotation of the bead in one direction would require energy, while rotation in the other would release energy. This was like going up and down a spiral staircase, only rather than working against gravity, it must work against an electric potential. Since this system was in a fluid, its small size made it subject to the random motions of molecules within the fluid. The bead would then randomly joggle back and forth, more often down the potential, but occasionally it would move up. Using a high-speed camera, the velocity of the bead could be determined, providing information to a computer. The computer converted this information into energy by selectively raising the voltage behind the bead when it was moving up the potential. This was like putting up a wall behind it, each time it moved ahead. Acting as the demon, the computer was able raise the potential energy of the system, without directly supplying any energy to it, through the use of information.

While the equivalence of information and energy is a fascinating concept on its own, it leads to some even more remarkable philosophical conclusions. If there exists a being who is truly all knowing, that being would necessarily be all powerful as well. This is a consequence of the equivalence of information and energy; infinite information would equate to infinite power. This demonstrates another characteristic of God. Being all knowing, requires that He also be all powerful. It also represents a mechanism that only God could use to supply the universe's low entropy condition at the beginning of time. The very special initial state of the universe may have resulted from knowledge contained within the mind of God.

Free Will

The question of free will has spawned debates among philosophers and theologians for millennia. Do we really possess free will? Or are we simply acting out our programming that originates in our DNA and is moderated by our experiences. This goes beyond nature versus nurture. Do we have any choice at all, or are we born predisposed to a particular venue? Though this topic is complex and touches many fields of study, there are reasonable solutions to this puzzle as we continue to consider the implications of the uncertainty principle.

In theology, the question of free will leads to the topic of predestination. That is, are the choices we make, especially the choice of following or not following God, the result of our own free will, or are they purely the Creator's discretion in making beings that are preprogrammed for good or for evil? This question takes a similar form in philosophy. Do the natural laws govern the mind as they do the motions of the planets? As was discussed earlier, the notion that all future events are predictable from current conditions, is called determinism. But determinism was found to be false due to the uncertainty principle. Within the realm of the very small, quantum effects dominate the behavior of atoms and their constituent particles. According to quantum mechanics, there is a degree of randomness at work within all interactions at microscopic scales. The activity of the mind involves chemical and electrical signals that occur at the quantum level and are therefore influenced by these random effects. There is no way to predict a human thought, even if you knew the function of every neuron in the brain.

Due to the uncertainty principle, our fate is not sealed. We are not simply acting out a program; DNA and experiences do not wholly determine our choices. Though these things do play a part, there is more to it than that. There is a seemingly random effect at the quantum level that is interacting with the logical program of behavior in the brain. This is our invitation to free will. The uncertainty principle masks the interaction of the soul. No experiment can isolate the soul because it acts below the limits of certainty. Under the cloak of randomness, the soul guides the function of the brain, bringing completion to the mind. Through this mechanism, the soul provides human beings free will.

When I first wrote this section, I was unaware of theoretical work in this area. A mechanism for these ideas has been proposed called "orchestrated objective reduction" (Orch-OR). This links consciousness in humans to quantum computations within the cellular fibers of the neurons in our brain.[7] Essentially, quantum computers are at the heart of each neuron's function, leading to incalculable decisions and choices. Quantum entanglement is critical to the Orch-OR theory and has been demonstrated to occur in many biological systems, lending credence to this notion. This quantum effect allows distant parts of the brain to interact as one functioning unit.

Evidence for Orch-OR can be found by examining the effects of anesthesia. Orch-OR theory postulates that the connective fibers (microtubules) within the neurons in your brain process and transmit information via entanglement of quantum states. Drugs that result in anesthesia have been shown to bind to the microtubules, disrupting quantum entanglement. This shuts down both quantum computation and transmission of information, leaving the subject unconscious.

This truly mind-blowing theory of how consciousness works exposes the mind's utter dependence on "random" quantum effects. It also describes the collapse of the quantum wave-function into a specific observable state as a "choice".[8] Not only in our minds, but throughout the universe as well. This view of quantum processing, directly implies the interaction of the soul in our mind, and by God throughout the universe. While the distinguished scientists who have developed this theory may have other metaphysical explanations, Orch-OR is most complete when we consider God and soul into the equation.

Attributes of God

In addition to the fine-tuning of the physical constants, the Orch-OR theory demonstrates how the exquisite fine-tuning of quantum behavior and of the properties that allow for quantum entanglement are required to allow for consciousness. Without quantum computing, there is no consciousness. God had this in mind when crafting these marvelous quantum laws.

By examining the universe and the laws of nature, we can begin to detect the traits of God. Great power and omnipotence can be seen by the scale of the universe. Wisdom and intelligence are required in crafting the details of the Big Bang and the fine-tuning of the laws of physics. It becomes apparent why God must be powerful enough to perform his creative duty, when considering the equivalence of information and energy. The very nature of an all-knowing God requires that He be all powerful. This tapping of infinite knowledge represents a mechanism which could produce the low entropy state that existed at the beginning of time. Only God has an infinite supply of information that could be utilized in this way. "The Lord by wisdom founded the earth; by understanding he established the heavens;" (Proverbs 3:19 ESV).

His purpose can be seen throughout creation: a long-lasting universe, ideally suited for sentient beings to thrive in. He made a way for them to be fully conscious, not simply as robots, but truly having choice.

We know that the Creator must be continuously involved in the universe, as implied by the uncertainty principle. If He were not, then no final outcome could be guaranteed. Yet this continuous involvement in the universe comes without having need to break any of the physical laws. He exercises control at the quantum level, with his actions masked by the unpredictable nature of that domain. For this reason, the world is discoverable to us through science and observation. By allowing natural law to flow as expected for human observers, God grants us the ability to understand nature and influence it for our own advantage.

Chapter Seven
The Universal Design

We have almost concluded our review of cosmological evidence, yet another interesting observation has been recognized. In the book, *The Privileged Planet*, compelling arguments are made for the nature of our universe to be intentionally discoverable. Here, a NASA scientist shows how the characteristics and locations of habitable planets happen to coincide with the best locations and conditions for astronomical observation.[1] This result cannot be explained by the multiverse argument, so it represents a new avenue of evidence for creation. A couple of examples of this are presented here.

Total Solar Eclipse

This starts with the fact that our moon is the perfect size and distance relative to our planet to support plate tectonics and maintain our rotation axis, necessary for the development of complex life. The Earth is at just the right distance from the sun, relative to the sun's size, to allow the mild temperatures required for life. These requirements for habitability perfectly coincide with the requirements for a planet to experience a total solar eclipse. A total solar eclipse can only occur on a body where there is a moon or planet that has the right combination of size and distance to match the apparent size in the sky as that body's sun. For the Earth-moon system, this correlation is better than 98%. It turns out that the best (and only) location in our solar system to view a total solar eclipse is on our planet Earth.[2] Not only is the Earth a good location to view an eclipse, but any planet in the habitable zone of its star with the right size moon to maintain plate tectonics will also be likely to experience total eclipses.

Other than being fascinating to behold, a solar eclipse is an important event for the advancement of science. It is through solar eclipses that we have learned a great deal about solar physics and even relativity. Einstein's theory of general relativity predicted that light from distant stars passing near the sun would be bent by the sun's gravity. But measuring this effect would have been impossible in his day, if it weren't for the ability of a solar eclipse to block the glare of the sun. A solar eclipse provided an early opportunity to test and verify his theory.[3] Otherwise we would have had to wait until spacecraft were available to test this prediction. But as it was, general relativity gained wide acceptance following these first tests.

The Galactic Habitable Zone

Our location in the galaxy is within what is regarded as the galactic habitable zone. Like the solar system's habitable zone, the galaxy's habitable zone represents a band of space where a habitable planet could reside. Too close to the galactic center, the intense radiation of that area would be fatal to the development of complex life. Too far out and heavy elements are too scarce.[4] These elements are required for building rocky planets with iron cores like that of the Earth. Places like globular clusters, where the stars are close together, would not allow for a planet to exist in a stable orbit for long due to frequent close encounters with other stars. So the habitable region of the galaxy is a fairly restricted zone.

It turns out that this same location is also ideal for observing the universe as well. At the galactic center, there is so much interstellar dust that observing the rest of the galaxy would be impossible at most wavelengths. In a globular cluster, the unfortunate planet's sky would be too bright for observing faint objects like galaxies and other globular clusters.

The properties of a breathable, life-supporting atmosphere also support discovery. Our oxygen-nitrogen atmosphere is transparent with only partial clouding so that observing the universe is quite easy. No other planet in our solar system with a thick atmosphere is also transparent enough for gazing into space.

Albert Einstein once said, "The most incomprehensible thing about the world is that it is at all comprehensible." He was baffled by the discoverability of the universe. The argument has been made (from the multiverse perspective) that we are in a universe that is friendly to life because we could only exist in such a universe.[5] But there is no natural reason that the means of discovery would also coincide with bio-friendly planets in a bio-friendly universe.

It is truly startling that these two derived properties of habitability and discoverability could happen to coincide just by chance. The most favorable places for complex life to emerge are also the best for observing the universe. However, this is not surprising if one has acknowledged that the universe was not only created but created for a purpose. By this axiom, we see how the Creator intended not only for intelligent life to thrive, but also to observe the heavens, discover the wonders of creation, and learn how it was brought to be.

Summary of Cosmological Evidence for a Creator

The question of God can be said to be a religious question. But the question, "How did the universe originate?" is scientific. But these questions are really the same. For if God exists, then He created the universe. Conversely, if the universe was not created by

God, then there must not be a creator. This is the very definition of God, the Creator of all that exists. If a scientist applies the modern assumption to the question of origin, then in so doing he neglects the most viable conclusion. The fact that so much evidence exists for a created universe means that the modern assumption (which proposes that there is no supernatural influence in the universe) must be laid aside to objectively address this question. Here is a brief summary of our findings so far.

The universe is temporary. An eternal universe might not require a creator. But we know that the universe had a beginning due to the verification of Big Bang theory. By the Second Law of Thermodynamics we know that the universe will eventually come to an end. Since the expansion of the universe is accelerating, the oscillating universe theory is proved invalid. Both of these facts have eliminated the possibility of an eternal universe. This leads to the first cause. Something outside our universe had to be responsible for its creation. This makes the existence of an intelligent creator a viable possibility.

The constants of nature controlling all natural phenomena are fine-tuned to a high level of precision. These physical constants, such as the gravitational constant, the mass of a proton, the charge of an electron, and the initial parameters of the Big Bang, such as average density, the cosmological constant, and many others are extremely critical to our existence. If any of these were only minutely different, our universe would be a very barren place, devoid of stars, galaxies, or life. These parameters appear to be selected expressly for the purpose of producing a richly complex universe, ideally suited for life.

Fine-tuning directly supports the existence of a creator, as this property would be expected of a created universe. This would be a natural prediction of an older argument for God that states, "the world is so orderly and perfectly suited for us because it was created for us." The naturalist might say, "No, we are simply evolved to suit our environment." But the universal constants of nature are indeed constant throughout time and space as far as we have been able to measure. So there exists no detectable mechanism to allow for environmental selection. We see that multiverse theories are contrived in an attempt to explain away the created nature of our universe. But lacking evidence, these ideas cannot detract from the fine-tuning argument.

The fact that multiverse theories could be entertained by so many in the scientific community, lacking credible evidence, shows the merit of fine-tuning. Seemingly, there is even a bit of desperation to find a counter argument since evidence for fine-tuning is so great and examples are so numerous. Alternate universes are the only naturalistic explanation. If there is no multiverse, there must be a creator. But as we have already discussed, the multiverse fails as a theory, it fails to explain fine-tuning, and it fails to explain the concurrence of discoverability and habitability in the universe.

One thing that the multiverse theories show us is that something outside of the physical universe is required to explain our existence. And that fact is recognized by many cosmologists. Even if other universes did exist, the theories behind them still need a beginning point, still require that nature have very specific fine-tuned properties, and would still require an explanation for their origin.

The uncertainty principle represents a limit to deterministic science. With the uncertainty principle, we observe the random effects at the subatomic level. This is not a limit in our ability to take measurements or in our understanding but is a fundamental property of quantum physics. So the universe has a built-in mechanism for control by the Creator that would be invisible, immeasurable, and untestable to us. This accounts for both gentle influences for continual guidance as well as more abrupt acts in the form of miracles. The uncertainty principle also provides for free will by masking the interaction of the soul.

Why would the Creator limit his actions to the quantum level? By doing so, He makes the universe orderly and comprehensible to humans. This allows for discovery and allows us to have power over our environment. It also allows for free will by giving us control of our own mind. A natural explanation does not imply that there is not a supernatural reason. The natural and supernatural are two sides of the same coin; one side does not exist without the other.

The initial low entropy state of the universe is hard to understand from a naturalistic perspective. There is one way to produce a low entropy state, without an external energy source—through the equivalence of information and energy. This provides a mechanism, that only God could employ, to create the low entropy state of the early universe. This question has long troubled cosmologists, but here we see that only God could create this low entropy condition, since only God has infinite knowledge and the ability to control quantum events.

The fact that science can predict so much natural phenomena has been a popular argument against God's existence. But with quantum mechanics we discover that there is a limit to science's ability to predict events at very small scales. So the limits of science are reached, and beyond this point science is incomplete in predicting reality. This boundary of nature opens the door to the supernatural.

The Simple Conclusion

The modern assumption of naturalism performs poorly when applied to cosmology. This assumption led to several failed predictions. The static universe theory and then the oscillating universe theory both proved false only after years of support in cosmology. These theories were in violation of the Second Law of Thermodynamics but were widely accepted due to their agreement with the modern assumption. They were not accepted because they were a best fit to the available data, but because they fit with ideology. This is a dangerous path for science. The multiverse is the latest product of the modern assumption. With no evidence or testable predictions, its support results from an ideology that in the past has led cosmology astray.

On every scale the modern assumption claims we are the result of a chance occurrence caused by random effects on infinite scales of time. But now science has probed the depths of the universe, and chance can no longer explain it. Luck as a creation mode has ran out. If we recognize the failure of the modern assumption and examine the data and theories that are well established, then there is one conclusion. The universe was created by an intelligent architect, who engineered it for a purpose.

The Creator's qualities are written on the very fabric of time and space. We observe wisdom and perfection in the precisely balanced equations that allow for stable atoms and for life. Omnipotence and active participation are implied by the uncertainty principle. The personal involvement of the Creator is not limited to the cosmological scale; the uncertainty principle allows for continual command of every atom without violating a single law of physics. Indeed, the acts of God are indistinguishable from the laws of nature. The purpose of our universe is to provide a place where intelligent creatures can thrive, creatures who can discover the wonders of creation and the Creator Himself.

Now we cannot be too vague as to the identity of this Creator. It is obvious to all, as the majority of the world population recognizes him as the God of the Bible. Jews, Protestants, Catholics, Orthodox Christians, and many other faiths call upon the same creator God. The differences between these world religions are great, but they all share belief in God and his role in creating and sustaining the universe.

Further Questions

By this point in the journey, you might think it was complete. Here cosmology has provided definitive evidence for the existence of God. Even some of His character qualities are evident in the design of the universe. But as is often the case, answers lead to questions. What about evolution? What about the vast scales of time in the Earth's history? Some would regard the details of biblical creation to be in contrast to the findings of natural history as read from the geological record. They would claim that this is evidence against the existence of a Creator or at least his identity as the God of the Bible.

There are some believers who might reject scientific findings in order to justify their understanding of biblical creation. They would argue that either the fossil record has been misinterpreted, or that the world was for some reason created to appear older than it really is. Before embarking on this quest for knowledge, there were times when I had entertained some of these thoughts. But I quickly realized the error in this reasoning. This position is difficult to justify due to the critical nature of scientific research. Dating techniques can be used on items of known age to validate the method. Due to the critical nature of science, it would be difficult for large errors in fossil ages to be commonplace.

For the next few chapters, we will review what science has discovered about the Earth's ancient past. How did major features of the Earth form? When did they form? How long did it take for the various forms of life to appear on Earth? Then we can compare these conclusions with the biblical creation account, to see if any claims of contradiction or agreement are valid. Are there any predictions in the creation account that can be tested? How old does the Bible say the Earth is? Through history, science has given us an ever changing picture of the most ancient events. With each generation we get a more complete understanding. While science has not reached perfect knowledge of the past (and will never), it is closer than it has ever been to being able to describe the details of Earth's origin and history. Many fascinating discoveries have been revealed only in our time. It will be well worth the study to determine if natural history agrees with our findings from cosmology as it pertains to God.

Chapter Eight
Natural History of the Planet Earth

The natural history of our planet is recorded in the geology of the rocks and minerals within it. Other solar system bodies also record events from the past, giving clues to the Earth's beginnings. With these clues, science provides a window to these ancient times.

A study of the heavens reveals a multitude of solar systems with worlds of their own. We see planetary systems in all stages of development, some still forming, some very ancient and nearing a certain end. Using what has been observed, scientists form theories and then test these models with computer simulations. While our understanding of our planet's beginnings is still developing, much has been learned. This provides us a glimpse at the Earth in its infancy.

Origin of the Elements

Shortly after the Big Bang, the universe consisted of evenly dispersed hydrogen, helium, and trace lithium gas.[1] Eventually, slight density variations were magnified by gravity as denser regions contracted, drawing in material from surrounding regions of lower density. The gases coalesced into large clouds, which were the first galaxies. Inside these proto-galaxies, the first stars formed. They quickly consumed their fuel converting the hydrogen and helium into heavier elements. Within a few million years, the most massive of these stars end as supernovas. This spreads the newly manufactured elements into the galactic medium.[2] Many generations of stars are born and die, further enriching the galaxies with heavy elements. The first stars had no planets. There was no rock or ice from which to form solid bodies. But later generations of stars would have increasing amounts of solid material available for planet building. Occasionally, the death of a star would occur near a cloud of gas and dust. The supernova explosion would send a shock wave into the cloud, compressing the nearby gas to a critical density. The cloud would begin to fragment at one end, forming new stars from the compressed gas.[3] Fragmentation would continue through the nebula until thousands of new stars form.

Planet Formation

When the universe was just over 9 billion years old, about two-thirds its present age, our solar system formed 30,000 light-years from the center of the Milky Way Galaxy.[4] [5] The solar nebula fractured off from a larger cloud of gas and dust into a swirling disk with the sun forming at the center. Heat from the new sun melted dust grains so that they began to stick together. Within the disk, dust would collect in denser regions to form proto-planets, numerous small bodies of ice and rock. These small bodies would frequently collide coalescing into larger planets. If a planet gained enough mass, it would gravitationally attract gases from the nebula. The larger bodies, such as Jupiter and Saturn, would acquire massive amounts gas, so that the gas would become the bulk of their final composition.[6] Closer in, the planets were smaller due to smaller feeding zones and competition with the sun for material. The Earth was frequently bombarded by incoming meteors throughout the accretion period that lasted up to 100 million years.[7] [8] The nebular gas was mostly hydrogen and helium so these gases made up most of Earth's initial atmosphere.[9]

It is believed that a Mars-sized planet struck the Earth partway through its formation. The impact sent large amounts of material from the crusts of both planets into space, while the impacter's core sunk into the Earth. The event would have melted a substantial portion of the Earth's surface creating a magma ocean. The material ejected into space soon coalesced under its own gravity to become the moon.[10][11] This theory explains the low density of the moon and why the moon's composition matches the composition of the Earth's crust, minus volatiles that would have escaped into space.

The Earth had acquired the majority of its mass by the end of accretion, about 4.45 billion years ago.[12] It had long been assumed that the Earth was very hot and possibly even molten during much of the first half billion years.[13] But recent evidence has revealed cooler conditions during this time. Zircons (ancient crystal grains that can be accurately dated) have been found dated up to 4.4 billion years old.[14] The ratio of certain isotopes within these crystals indicate the conditions under which they formed. Not only did these crystals form in a cool (< 200 °C or 392 °F) environment, but also in the presence of liquid water. This suggests that not only did the Earth have a solid crust by about 4.4 billion years ago, but oceans had already formed.[15]

Geologic Time: First Three Eons

EON	ERA	PERIOD	AGE*
Hadean			4550
			4000
Archean	Eoarchean		3600
	Paleoarchean		3200
	Mesoarchean		2800
	Neoarchean		2500
Proterozoic	Paleoproterozoic	Siderian	2300
		Rhyancian	2050
		Orosirian	1800
		Statherian	1600
	Mesoproterozoic	Calymmian	1400
		Ectasian	1200
		Sterian	1000
	Neoproterozoic	Tonian	850
		Cryogenian	630
		Ediacaran	542

*Boundary age in millions of years ago.

Development of the Crust and Atmosphere

Early conditions with liquid water may imply an environment similar to the present, but to the contrary, it was quite an inhospitable place. The temperatures were likely in excess of 100 °C (212 °F). The high atmospheric pressure of that time would raise the boiling point of water, allowing it to be liquid even at temperatures above 100 °C (212 °F).[16] This is the same effect as within a pressure cooker. When the Earth first solidified, it would have been much smoother. And though volcanism would have been highly active, it would take time for any variations in elevation to accumulate on the Earth's surface. For a smooth Earth, its current volume of water would cover it four kilometers (2.5 miles) deep.[17] This water world may well have been devoid of land for many millions of years.

The early Earth was likely very dark as well. At this time, the young sun's intensity would have been only 70% of its current level.[18] Though we do not have a lot of direct evidence as to the composition of the atmosphere at these early times, there are a number of theoretical models that predict some of its characteristics. The atmosphere may have been hundreds of times denser than the present, having massive amounts of CO_2 and similar cloud cover as the current atmosphere of Venus.[19] The greenhouse effect of a thick carbon dioxide atmosphere[20] likely made up for the reduced solar luminosity along with increased heat from radioactive elements in the Earth's core. Though the Earth would have been quite hot by our standards, the large amounts of water vapor, methane,[21] sulfur dioxide,[22] hydrogen,[23] and other gases would have contributed to dense clouds and hazes that significantly darkened the surface. This can be seen on Venus today, with only a small fraction of the sunlight making it to the ground. The Earth is farther from the Sun than Venus, and the sun was less luminous at this time. These two effects alone would reduce the radiance at the Earth's surface to only 37% of modern Venus for matching atmospheric composition and density.[24] Add to this, the heavy volcanic activity, rich hydrocarbon chemistry, and much higher water content, then the early Earth would have been a dark place indeed.

Covered in miles of water, and dark, the early Earth would have appeared very similar to the fictional planet Kamino from the movie,

Star Wars Episode II: Attack of the Clones. This is the planet that Obi Wan Kenobi visited in his search for Senator Amidala's attacker. This dark stormy water world provides a picture of what the early Earth must have been like. Only it was warmer, and completely devoid of life. Frequent storms likely ravaged the surface with heavy rains and lightning.

Over time, the surface temperature cooled further, allowing for less water in the atmosphere and less cloud cover. Hydrogen in the atmosphere would escape into space due to action by the solar wind. By 4.0 billion years ago, the cores of the modern continents (cratons) began forming.[25] The cratons may have still been under water at this time, but would eventually be pushed upwards by the action of plate tectonics. When their peaks exceeded the sea level, dry land appeared. These may have been small islands at first surrounded by large areas of shallow water. Crust volume increased very slowly for the first 1.5 billion years of Earth history.[26]

During the Apollo program, rocks had been collected from many different locations on the moon. Examination of meteor impact melts indicate a large surge of impacts at about 3.9 billion years ago. This came after a long period of relatively few impacts from 4.4 to 4.0 billion years ago.[27] If the moon received a large dose of meteors at this time, then the Earth would have endured the same fate. This has become known as the Late Heavy Bombardment. It is difficult to determine the impact rate and the size of the impactors so far back in history, but this event may have been sufficient to temporarily vaporize the Earth's oceans.

The oldest sedimentary rocks found are 3.8 billion years old.[28] At 3.7 billion years ago there are minerals that appear to be altered by microorganisms. However, there are non-biological explanations for the find, so this evidence is somewhat inconclusive. But by 3.5 billion years ago the presence of life is more certain with the existence of stromatolites and fossil microbes.[29] Stromatolites are layered formations made by colonies of single-celled organisms in shallow water. From 3.5 to 3.0 billion years ago the Earth's climate was mild and warm.[30] Photosynthesis in microorganisms originated no later than 2.8 billion years ago and possibly as early as 3.7 billion years ago.[31]

Life has existed on Earth for billions of years. While there is clear evidence for its existence so early in history, scientists have very little understanding for how life began. This topic will be explored in detail in chapter twelve. Microbial life flourished, but multicellular life would have to wait several billion years for its turn to come.

For the first 1.5 billion years, there was little to no land. Our once dark water world was now a more temperate home as sunlight filtered through the clouds to illuminate the liquid surface. Mild temperatures and a world of ocean might make this epoch sound inviting to sea loving individuals. But our atmosphere still lacked oxygen, making it poison to most forms of current life.

By 3.0 billion years ago, only 20% of the Earth's crust had formed.[32] This thin crust was predominately under water. Then from 3.0 billion to 2.0 billion years ago, the Earth went through a process of extensive continent building. It was during this time that most of the continental mass was formed.[33] By the end of this period, over half of the current land area would be present. For the last 2 billion years continent buildup has continued at a much slower rate until the present day.[34] Once significant land mass exists, continental weathering reduces carbon dioxide through reactions with minerals and flowing water.[35] This process releases oxygen into the environment, even without photosynthesis from plants or microorganisms. Besides weathering, there were microorganisms involved in producing oxygen as well. However, neither source of oxygen would make it into the atmosphere in significant quantities, since it quickly reacted with natural reducing agents that existed at the time.

The first known glaciation event occurred about 2.9 billion years ago.[36] There may have been earlier periods of ice or local occurrences at the poles, but any evidence of these earlier freezes has been erased by time. Ice ages are thought to be caused by reductions of greenhouse gases in the atmosphere, either due to weathering of rocks or by the action of photosynthetic organisms.

For hundreds of millions of years, the Earth enjoyed a warm climate from 2.8 to 2.5 billion years ago. The land would have had beaches and mountains but would have lacked any form of plant or animal life. Then there began a period of frequent ice ages between 2.45 and 2.22 billion years ago. During these glaciations, a significant fraction of the oceans would have been covered in ice. The time of these ice ages corresponds with what is known as the "Great Oxygenation Event".[37] At 2.45 billion years ago oxygen was scarce (less than .0002%) in the atmosphere.[38] But by 2.0 billion years ago oxygen was a significant atmospheric component at about 3%.[39] Oxygen levels rose at that time to the highest level the world had yet seen. This caused major changes in the chemistry of the land, oceans, and atmosphere. The increase in oxygen would have caused methane and carbon dioxide to be reduced, both of which are greenhouse gases. In the geological record, banded iron formations were common up to the point of the oxygenation event. As oxygen levels rose in the oceans, iron would have oxidized and precipitated out. Once all the iron dissolved in ocean waters was oxidized, no new formations would be deposited. These formations provide a clear indicator of oxygen presence and the timing of its domination of ocean chemistry.

A cyclic process of greenhouse gas buildup, then oxygenation would repeat in lockstep with the ice ages. As photosynthesis by microbes and continental weathering would increase oxygen and reduce carbon dioxide and other greenhouse gases, the temperatures would drop. This would send the Earth into an ice age. The more land and sea that became covered in ice, the less photosynthesis and continental weathering would occur. This allowed volcanic emissions and other processes to add greenhouse gases back into the atmosphere, warming the planet once again.[40] This process would repeat many times through history.

The Earth's land masses had long been barren. The first evidence of biological activity on ancient soil is found at 2.6 billion years ago. The greening of the Earth's landscapes could have began by 2.2 billion years ago. Urn shaped micro-fossils up to two millimeters long have been discovered in ancient soils from this time period.[41] These fossils represent lichenized prokaryotes or possibly an early eukaryote. In either case, this would have dramatically altered the prehistoric scenery and the chemistry of the soil.

For a billion years the Earth was quiet with few major changes or notable events. The time period from 1.8 to 0.8 billion years ago is known to geologists as the "Boring Billion". The climate was warm with no known ice ages. The oxygen level in the atmosphere was approximately constant at about 3% during this time.[42] Very few fossils exist from this period, which consisted only of altered minerals and obscure microbial remains. It seems that whatever form of life existed in this time period was in stable equilibrium with the environment for a billion years.

The first 3.7 billion years of Earth history lacked plants, animals, or any other form of complex life. Until 800 million years ago, small organisms ruled the Earth. With the geological stage set, the Earth waited quietly. Meanwhile a revolution was brewing, culminating in the arrival of macroscopic life.

Chapter Nine
Life history on Earth

Locked within the earth are recordings of life's past forms. Seasonal flooding, winds, ocean currents, or other natural process will often deposit materials in layers. These layers make up the strata. The strata consists of layered rock, fossils, and minerals, which have been preserved to the present day. In addition to seasonal buildup in the strata, there are layers caused by catastrophic events, such as volcanic eruptions, meteor impacts, wild fires, etc. Each layer tells a story of the age from which it formed.

Reading the strata is difficult due to variable conditions throughout time, but a number of techniques can be employed to make reliable correlations. One way is through the observance of catastrophic events that leave markers in the strata worldwide. By dating these events and recognizing their signatures in the strata, a minimum and maximum date can be obtained for the layers in between. Often radioactive minerals can be found that provide atomic clocks for dating the layers that they originated in. The past chemical composition of the atmosphere and oceans also leave their mark in the strata. The properties of the major layers are recognizable worldwide and are named according to the geologic period they originated in. The US Geological Survey has published the official names of the layers and determined the age at each layer's boundaries.[1] This provides an effective way to estimate the age of specimens found within the strata.

Geologic Time: The Current Eon			
EON	ERA	PERIOD	AGE*
Phanerozoic	Paleozoic	Cambrian	542.0
		Ordovician	488.3
		Silurian	443.7
		Devonian	416.0
		Carboniferous	359.2
		Permian	299.0
	Mesozoic	Triassic	251.0
		Jurassic	199.6
		Cretaceous	145.5
	Cenozoic	Paleogene	65.5
		Neogene	23.0
		Quaternary	1.8
			0.0

*Boundary age in millions of years ago.

Geological time is divided into eons, eons into eras, and eras into periods. The period boundaries are chosen for the identifiable features within them. Each period is distinct in the types of fossils it contains, both for plant and animal remains.

The story of life's beginnings is intermixed with our planet's development. In the last chapter we saw that life, in microbial form, has existed on Earth since early in its history. Life has influenced the development of the Earth, altering it over time. At 1.85 billion years ago, the Earth had continents and oceans, weather and seasons, microbial life in vast varieties, a mildly oxygenated atmosphere, but still lacked macroscopic life.

Photosynthesis and Precambrian Life

For about a billion years, oxygen levels remained unchanged at approximately 3% of atmospheric composition from 1.85 to 0.85 billion years ago.[2] Whatever processes that maintained this oxygen level would also have remained constant as well. An oxygen atmosphere is not stable without some mechanism for maintaining it. The most well-supported evidence points to microbes capable of photosynthesis as the principal source of oxygen during this time, not only in the oceans, but on land as well. This suggests that these microorganisms remained relatively unchanged for at least a billion years. Clearly, whatever new species or variant that did appear during this time had no significant impact on the Earth's atmosphere.

But then things start to get interesting at 0.85 billion years ago when oxygen began to increase at a steady pace for about 300 million years, bringing oxygen up to about 10% to 15% of atmospheric composition by 540 million years ago.[3] On the scale of Earth history, this was a revolutionary event. Something changed in the biosphere around 850 million years ago that allowed for the accumulation of large amounts of oxygen in the atmosphere. Most likely this was caused by some novel life-form, perhaps the appearance of macroscopic photosynthetic organisms.

Many primitive species may have played a part in the oxygenation of the atmosphere. Fossil evidence for lichens have been found at the Doushantuo Formation in southern China that are dated at 599 million years old.[4] Fossil liverworts on land may have been present as early as the Cambrian, but these fossils are of poor quality and badly degraded.[5] Small land plants decompose quickly, so fossilization is rare or non-existent for the earliest plants. Algae has been present for up to 1.9 billion years,[6] so it is unlikely to have been responsible for the rise in oxygen levels during this time. Algae fossils on land have been found dating 1.2 billion years old.[7]

Other forms of life show up within this time period that may have had an effect. DNA evidence suggests that fungi and lichens may have colonized land as early as 900 million years ago.[8] Researchers can track molecular changes in common proteins that can be compared between species. Knowing the average rate of mutation, which is calibrated by reference points known from the fossil record, the approximate time of divergence between two species can be determined. This method has provided an estimate for the divergence of moss and vascular land plants at about 703 million years ago. This is reported as a minimum age for the colonization of plants on land.[9] Since fossilization is only a chance occurrence, these forms and others may have been present earlier than the fossil record indicates as suggested by DNA studies. This evidence for early land plants corresponds well with the oxygen level increases in the atmosphere suggesting that primitive land plants may have been a principal driver in the accumulation of oxygen from 850 to 540 million years ago. This time was a time of transformation as the land became greener and the atmosphere changed dramatically.

At 540 million years ago, oxygen had reached about 15% of atmospheric composition compared to 21% today.[10] The other constituents of the atmosphere were also similar, though carbon dioxide levels were higher. If you had been there, it is likely that you could have breathed the air of that time. You would have looked up into the sky to see much of what you see today, a bright yellow sun, blue sky, and partial or occasional clouds. At night the moon and stars would be visible. The lack of artificial light and air pollution would have offered spectacular views of the Milkyway and other deep sky objects.

Before 540 million years ago, the record of animal life is sparse but consists of a few small soft bodied life-forms. These simple aquatic organisms are known as the Ediacaran Fauna named for the geological period in which they lived (Ediacaran Period: 630 – 542 million years ago).[11] They were strange and unlike life today or even what existed in the adjacent geological periods. Many of these organisms lack a modern counterpart and cannot be identified as plant or animal. Though some were believed to be green and rooted to the ocean floor and were quite plantlike in appearance.[12]

The Cambrian Explosion: Aquatic Animals

The fossil record changes abruptly at 542 million years ago, with the appearance of many diverse forms of shelled aquatic animals. This is known as the Cambrian Explosion. Any detailed study of the fossil record or of evolutionary history will reveal that prehistory is punctuated with many episodes characterized by explosions of new species that appear in very short periods of time. These explosions are often separated by millions of years where there is very little biological change.

The Cambrian explosion marks the beginning of the Cambrian Period. The Cambrian Period is defined by the US Geological Survey as 542.0 to 488.3 million years ago.[13] Animal life made its debut in the Cambrian,

100 cm (2.5 inches)

Trilobite Fossil.
Trilobites thrived in Earth's oceans from 520 to 250 million years ago.

though some primitive animal forms are believed to have originated earlier. To be sure, it is in this period that animals gained vast diversity and rose to dominate the fossil record. Animals represent just one of the kingdoms of life. Life is categorized from the highest level by domain, then kingdom, phylum, class, order, family, genus, and species.[14] This organization structure groups similar organisms according to their physiology and their placement in the evolutionary tree of life.

During the Cambrian Period nearly all phyla of animals appear in the fossil record.[15] These are small and simple forms but existed in great diversity. Each basic body plan, which makes up the general organization of essential functions and structure, is represented by a single phylum. These body plans all originated in this period and continue to be used by all animals to this day. It also should be pointed out that all animal life was confined to the oceans at this time. Chordata, the phylum to which we belong and includes all vertebrates, were represented by primitive jawless fish.[16] The Cambrian seas were filled with a multitude of aquatic animals. The remains of these creatures fill the Cambrian layer to the extent that these fossils are its most predominant feature.

Darwin saw the Cambrian Explosion as the single biggest problem for his theory of evolution.[17] The abrupt appearance of fossils in the strata is striking. There are many ideas to explain this event, but mysteries surround it even to this day. Almost every animal phylum was represented during this time and had been non-existent before. This seems to run contrary to the many long intervals within the fossil record when evolutionary change appears more gradual.

The significance of this period is not only limited to the phyla of animals that it hosted but also the many novel features these organisms possessed. Eyes, spinal cord, shells, scales, legs, brain, nerves, heart, blood, stomach, muscles, tail, eggs, sperm, and many other biological components originated with the organisms of the

Cambrian. These creatures could see, smell, hear, taste, balance, and feel.[18] The life-forms in the Cambrian were small and primitive but had nearly all of the basic elements that animals have today.

Post-Cambrian

In the following geological periods the diversity of oceanic life continues to increase at a gradual pace. In the Ordovician Period (488 – 444 million years ago) we find the first vertebrates with bones. The period is ended by an extinction event that decimates half of the world's species.[19] The Silurian Period (444 – 416 million years ago)[20] welcomes jawed fish, diverse forms of jawless fish, and sharks into existence by 420 million years ago. At this time, the land is only sparsely inhabited with a few small animals such as scorpions. Land plant spores show up in the fossil record 440 million years ago.[21] Though DNA evidence cited earlier indicates that plants likely arrived earlier than the oldest surviving fossils would suggest.

The world during this time was quite a different place. With oceans covering most of the Northern hemisphere, the continents were arranged closely in the South. Temperatures were generally much warmer than today, punctuated by intermittent glacial episodes on cycles lasting millions of years.[22]

The Devonian Period (416 – 359 million years ago)[23] is sometimes called the "Age of Fish" for its great abundance of fish varieties. Many new plant species covered the land. Plants grew ever taller in their competition for sunlight. Progymnosperms (tree-like plants) spread worldwide by 385 million years ago. These were the first wood-bearing trees with leaves.[24] Amphibians, mainly salamander-like creatures, show up in the fossil record during this period when few other creatures could survive on land. Then 365 million years ago, a mass extinction wiped out up to 70% of all species, near the end of the Devonian Period.[25]

Winged insects, reptiles, and crocodile-like synapsids inhabit

the land during the Carboniferous (359 – 299 million years ago).[26]
[27] Massive floras cover the Earth. Giant ferns, giant horsetails, peat
bogs, and strange clubmoss trees thrive. The bulk of the present day
coal deposits come from the massive swamplands of the Carboniferous. Wonderfully preserved plant fossils from this period, embossed
in the ceilings of many coal mines, are revealed when the coal is
removed. During this period, carbon in the atmosphere drops dramatically, and glaciers extend down from the poles. Many limestone
deposits are produced from dead sea creatures whose shells sink to
the bottom of the ocean.[28] These deposits eventually make their way
above water as tectonic plate movement drives the ocean floor
upward.

In the Permian (299 – 251 million years ago), there are reptiles
that begin to look more dinosaur-like, though smaller.[29] [30] Yet no
dinosaurs exist at this time. The Pangea supercontinent contains
most of the Earth's landmass. Large tropical forests flourished until
the end of the Permian, which is marked by the largest mass extinction event the world has ever known. Fifty percent of plants, nine-
ty-five percent of sea animals, and seventy percent of land animals
were eradicated.[31] The cause of this extinction is not firmly understood, but may have resulted from extensive volcanism in what is
now Siberia. This produced massive lava flows and ash plumes that
continued for thousands of years.[32]

The archosaurs, turtles, and flies originate in the Triassic (251 –
199 million years ago). This period also included the first dinosaurs
and crocodilians, and small primitive mammal-like creatures.[33] The
archosaurs ruled this age. Their large size and ability to cope with
varied environments gave them the advantage while life was still
recovering from the Permian extinction.[34]

The Jurassic (199 – 145 million years ago) and Cretaceous (145
– 65 million years ago) periods are quite famous, since during this
time dinosaurs dominated the Earth.[35] Pangea begins to break up

near the end of the Triassic, and the continents continue to separate through the Jurassic as well.[36] Birds appear during this time, themselves a clade of dinosaur, and are the only dinosaurs to have surviving descendants to the present day. Birds are classified as theropods, along with the tyrannosaurus rex and velociraptor among others.[37]

Primitive mammalian creatures, resembling shrews, were present during the time of the dinosaurs. They laid eggs, unlike modern placental mammals who birth their young. These creatures played a minor role in Cretaceous ecosystems, representing nothing more than a light snack for reptilian carnivores of that day. But the world changed abruptly at the end of the Cretaceous period, when a massive meteor struck the Yucatan Peninsula, creating the Chicxulub crater sixty-five million years ago. This event may have so disrupted the global environment that it lead to the Cretaceous-Paleogene Extinction and the annihilation of the dinosaurs.[38]

Empty ecological niches, long dominated by the dinosaurs, were now available for new species to fill. Placental mammals emerge soon after the Cretaceous-Paleogene Extinction.[39] An explosion of new species appears as the age of mammals dawns, resulting in the vast diversity of mammals large and small.[40] Most mammal species alive today have existed for millions of years. Animals such as bats, rhinos, camels, whales, deer, dogs, rabbits, pigs, and cats originate in the Paleogene (65 – 23 million years ago).[41] [42] Giraffes, kangaroos, apes, and hippos have their origin in the Neogene (23 – 1.8 million years ago).[43] [44] Many other animals exist in these periods that have survived to the present, especially the major families of placental mammals.

Fossil evidence shows the existence of many primitive human-like creatures of the genus homo, existing for the last two million years.[45] The last of these primitives, neanderthals, became extinct about 30,000 years ago.[46] Many of these species made use of stone tools. Chimpanzees, raccoons, bears, and other animals are also known to use rocks or sticks to accomplish a task. Many animals build intricate nests, dams, or other structures. So it is not surprising that these primitive homo species were able to do the same. To their credit, the tools they used were more sophisticated than other animal tools. The handax was a very specific and widespread tool that looked like a sharp triangular rock. It was chiseled out of stone and present at the archeological sites of many homo species throughout the last million years.[47] However, these primitives show no evidence of symbolic thought, nor did they practice religion or create art.

Anatomically modern humans are a very recent occurrence, appearing within the last 200,000 years. But for much of their existence, the artifacts they produced were quite similar to those of their predecessors. A few new inventions appear about 40,000 years ago, such as simple arrowheads, bone tools, and simple jewelry. Then there comes a revolutionary change in human behavior. Detailed and complex artifacts of human creation show up by 28,000 years ago. The sculptures, cave drawings, ornamental carvings, weapons, and elaborate jewelry became commonplace after this time.[48] These behaviorally modern humans would be quite indistinguishable from people today in physical appearance and mental capacity. Their skill, intelligence, and creativity was much like ours, judging by the artifacts they made. They possessed language, practiced religion, created art, and used symbolic representation. Behaviorally modern humans are extremely recent in the history of life on Earth, with intelligence and creative capacity appearing rather abruptly in the archaeological record.

Natural History vs. Cosmology

The study of Cosmology and the laws of physics point to the existence of an intelligent Creator, One who designed the universe with the purpose of making a place for intelligent beings like us to exist. But when we look at the natural history of life on Earth, it might seem so impersonal and unintentional. Hardly matching the impression we get when seeing the order and intentional nature of cosmology. Are we getting a contradictory verdict from cosmology and natural history? What can the findings of natural history say about the character or even identity of God?

The book of Genesis is one of the most ancient texts still in wide circulation. It is believed by many religions to be the revelation of our beginnings. As part of the Holy Bible, Genesis shares its place as the number one best seller of all time. Over six billion copies have been printed in modern times. For these reasons, the book of Genesis and other biblical texts regarding creation make a great place to start in the search for the Creator. If we want to be certain of the identity of this Creator, then we can analyze the descriptions of the creative acts as recorded in the Bible. When we examine scripture, do we find anything at all that foretells the findings of modern science? Or are these ancient accounts contradictory with the science of our day?

Indeed, the description of creation that I was taught in Sunday school is far from the picture of the beginnings that science paints. Just as competing scientific theories are numerous and must be subject to trial and testing, there are also competing viewpoints on the interpretation of the creation narrative found in Genesis. Because these scriptures are so ancient, we must take care in understanding the history and context of the narrative. Writing style and frame of reference must be considered to understand it fully. We cannot grasp any written word without knowing the context. Without realizing it, we use our past experiences and prejudices to understand all written work. I now know that the specific viewpoint I was taught as a child is based on the belief in a young Earth.

The young earth creation model draws its support from very devout followers, who accept this interpretation of Genesis on faith and generally do not consider any external sources as relevant. For this reason, this model rejects many findings of modern science in favor of tradition and the work of James Usher. In the year 1650,

Usher dated the creation of the heavens and Earth as Oct 23, 4004 BC.[49] His viewpoint is widely embraced for its simplicity and literal approach to interpreting Genesis. Though modern young earth creationists generally accept his findings on faith alone, he arrived at his conclusions through extensive historical study and making use of all of the scientific information available in his day. The fact that he relied heavily on science is forgotten by many in our time. Much of the science that was available then, however, is now very out of date.

While there are many other models for understanding Genesis, it is this interpretation which is often cited by opponents of creationism because it is so easy to disprove. But this is an unfair criticism; you wouldn't take the worst scientific hypothesis and by it judge all of science.

On the opposite end of the spectrum are those who believe the creation account of Genesis to be allegory, a purely metaphorical or spiritual description of creation. This viewpoint is popular among theistic scientists and some science-minded individuals. While this may be an easy out for any apparent conflicts with science, it misses the opportunity to offer another avenue of proof for the existence and identity of God, as well as verifying the God-inspired nature of the Bible. It also represents a mode of biblical interpretation that could be dangerous to the integrity of the scriptures.

But just as there have been many scientific models of the universe which have failed when put to testing and new discovery, some interpretations of Genesis fail when compared with geological finds, scientific data, the proper historical perspective, and correlation with other biblical passages. While there is more than one way to interpret the Genesis creation account, it will be shown in the next chapter why it is old earth creationism in the form of the day-age model that is supported in this book. This is both for its fit with scripture, its logical consistency, and its compatibility with scientific discovery of our time.

For much of the last century, science has taken a meandering path to discovering key facts about the origins of our universe, our world, and of life. This has been a transitional time for science, going from being unable to answer questions about origin, to now having a very firm grip on many aspects of universal history. As our scientific understanding has improved, so has the evidence for the existence of God.

If you have previously thought that science is in contradiction to the Bible, it is no doubt due to the popularizing of conflicts with out-of-date science or very narrow interpretations of Genesis. For those who are as yet uncertain about their belief in God or the Bible, it would be useful to see if the creation account of Genesis can be correlated to modern science. For believers, correlating science and the Bible can lead to a greater understanding of scripture. We can also gain insight into how God interacts with the world. Of course, we must remember that we have neither total understanding of the scriptures nor complete knowledge of science. So it is within this framework that we must look at general truths and concepts and not picking at every detail. The book of Genesis was written about 3500 years ago, and the author may have utilized source documents that were much older.[50] It was intended to be understandable to all people throughout history, through most of which, there would not have existed words to describe elaborate scientific statements of origin. Nor would these things be understood even if words could be found. But general statements about our world's origin could be made in simple terms that would be understandable to all generations. This would still be factual and testable in the later times when scientific knowledge would be available. Such tests could then determine if these ancient texts are indeed God-inspired by comparing the descriptions to the findings of science.

When you read the Bible's creation story, the first thing that might impress you is that it is very short. I have a small print Bible that is about 3.5 inches wide by 5.5 inches tall. Out of the 1389 pages of biblical text, the creation account fits within the first two pages. While creation is described in other areas of the Bible, this section contains the most complete and detailed account. To summarize the origin of the universe and our world in two small pages would be a remarkable feat, regardless of your belief about how it happened. In the next chapter we will analyze this biblical account of origin and compare it to natural history. The validation of biblical predictions about Earth's prehistory would constitute the best evidence yet for the existence of God.

Chapter Ten
Old Earth Creationism

Old earth creationism is a belief held by many that God created the universe and the Earth, but did so over the course of billions of years. There are a few different flavors of this belief, differing mainly as to where the billions of years fit into the creation account given in the Bible. One common approach is the day-age view. Of the various views on creation, I find this to be the most consistent with both science and scripture. The following analysis will examine the scriptures from this viewpoint.

The scriptures include both physical and spiritual perspectives, but science reveals only the physical condition. So any comparison with science must focus on the physical aspects, since science doesn't deal with the spiritual. Any predictions within the biblical creation account could serve as a test for the validity of the day-age model and as a testament to the God-inspired nature of scripture.

Imagine if God wished to describe the origin of the Earth and its transformation into a life-sustaining state. How would these things be described to people with no knowledge of science? We have already found that God can use natural phenomena to carry out his will. For this reason, some use of metaphor might be required for difficult concepts.

This would be especially true for the immense scales of time involved in the history of the Earth. To describe to ancient peoples processes spanning billions of years would have only confused them. For this reason, the use of the word "day" to describe a very long period of time is appropriate. In Hebrew, the word for "day" can also mean "age" or an indefinite period of time.[1] Even in English, the phrase, "In Abraham's day...", does not refer to a twenty-four hour period but indicates the age in which he lived. All of the days of the creation week began before humans existed, so the case could be made that these were days from God's perspective, not from man's. This speaks of God's timeless nature; all of the Earth's existence is as a week to Him.

Creation of the Universe

In the beginning God created the heavens and the earth. (Gen 1:1 NKJV)

The earth was without form and void; and darkness was on the face of the deep. And the Spirit of God was hovering over the face of the waters. (Gen 1:2 NKJV)

Gen 1:1 has been interpreted in two ways. One would be that it is a summary of the six days of creation described in verses three through thirty-one.[2] The other is to regard it as a separate and more ancient creative act.[3] Notice the description of the Earth in verse two. Clearly, the Earth exists, matter exists, and time exists in verse two, before the creative acts of the first week begin. Therefore, to consider verse one as merely a summary of verses three through thirty-one is to assume that Genesis leaves the creation of matter unexplained. (If these things exist in verse two, then they are not created in the acts of verses three through thirty-one.)

For by Him, all things were created that are in heaven and that are on earth, visible and invisible, whether thrones or dominions or principalities or powers. All things were created through Him and for Him. (Col 1:16 NKJV)

The Bible declares that God created everything, not simply the objects themselves, but the matter that composes them also. No believer would doubt that God created all the matter in the universe, so it is fitting that this original creative act would be proclaimed in the opening verse of the Bible.

Now we could ask, "When did this original creation of matter occur?" It either happened on the first day of the "six days" or it happened at some earlier time. While this question may seem trivial, any attempt to derive an age of the Earth from scripture is dependent on its answer. The Genesis creation account proceeds in a very structured way, following a specific pattern for each day. Each day begins with the phrase, "Then God said...", and ends with, "And evening and morning were...". Another pattern is found in the descriptions: everything that God creates during this week is proclaimed to be good. Verses one and two do not fit into either of these patterns. Verse two proclaims the Earth "formless and void". Indeed, verse three is the logical starting point for day one of the creation week. This fact alone eliminates any chance of dating the Earth from biblical genealogies.

Other clues to the original creation are given in Job 38:4-11 and Proverbs 8:22-31.

While as yet He had not made the earth or the fields, Or the primal dust of the world. When He prepared the heavens, I was there, ... (Prov 8:26-27 NKJV)

Proverbs declares that God created the universe before creating the Earth. He initiated and guided the formation of the universe long before the dust of the Earth ever existed. This is an excellent picture of the early universe that cosmologists describe, where the first generations of stars produced the elements which would be expelled as dust. This primordial dust would provide the necessary material for planet formation in subsequent generations of stars.

The true intent of Gen 1:1 is to describe the original creative act that occurred before the creation week of Gen 1:3-31. The "heavens and the Earth" in this passage is a phrase that means "all of reality" or "the entire universe".[4] The first verse of Genesis is in excellent agreement with the Big Bang model. The universe is not eternal but, in fact, had a beginning. The universe is also described as predating the Earth as in Big Bang theory. There is also support for Big Bang theory that is found in scripture elsewhere in the Bible. Here are a few examples.

And you forgot the Lord your Maker, Who stretched out the heavens And laid the foundations of the earth;... (Is 51:13 NKJV)

Thus says God the Lord, Who created the heavens and stretched them out,... (Is 42:5 NKJV)

...And has stretched out the heavens at His discretion. (Jer 10:12 NKJV)

And many other scriptures describe God stretching the universe. This alludes to the expansion of the universe, which continues to this day. Big bang theory describes the universe as expanding from a very compact initial state, just as we find here. Now let's look again at the second verse of the creation passage in Genesis.

The earth was without form, and void; and darkness was on the face of the deep. And the Spirit of God was hovering over the face of the waters. (Gen 1:2 NKJV)

Now that the original creation is understood, it becomes clearer what happened next. God surveys his universe and finds a water world, barren and dark. He selects the Earth to transform it into a marvelous creation, full of life. The Earth, Sun, Moon, planets, and stars of our galaxy were already formed by God's earlier work.[5]

The description of this primeval Earth found in Genesis 1:2 is verified by science. These conditions of the early Earth were described in detail in chapter eight. Modern planet formation models predict a thick atmosphere that would have shrouded our planet in darkness. Also, ancient zircons dated from 4.0 - 4.4 billion years old contain geological information about the presence of water in Earth's ancient past. There is much evidence that our world was once completely covered in water and devoid of continents for an extended time. So we observe in the Genesis descriptions a picture of the early Earth that has only been recently confirmed by scientists in this century.

The key to understanding the creation week is to realize the vantage point of the descriptions. The Genesis account is from the vantage point of the surface of the Earth. This is evident in Gen 1:2, as God's presence is described at our planet's surface. Also the term "earth" in the Bible often refers to land or the surface of the planet Earth. This vantage point is used to describe what a man would see if he witnessed the events of creation as they unfold. It also shows God's personal involvement on the Earth, not acting from far away.

Days One (Gen 1:3-5) and Two (Gen 1:6-8): Atmospheric Change

Then God said, "Let there be light"; and there was light. And God saw the light, that it was good... (Gen 1:3-4 NKJV)

From the vantage point of the Earth's surface, the thick early atmosphere became thin enough to be translucent, and light from the sun reached the Earth's surface. The light of the sun already existed, but it did not reach the surface of the Earth until the atmosphere and cloud cover became less dense.[6] This can be easily compared to the transformations of the Earth's early atmosphere as described by scientists. Much of the initial bulk of the atmosphere was lost into space early in the Earth's history. The early atmosphere contained large amounts of hydrogen and hydrogen containing compounds which were stripped away by the solar wind by about 4 billion years ago.

Then God said, "Let there be a firmament in the midst of the waters, and let it divide the waters from the waters." (Gen 1:6 NKJV)

The fog lifts from the surface, and the cloud cover is reduced as temperatures drop.[7] Planet formation models predict that the Earth would have gone through a phase of gradual cooling. As it cooled, evaporation from the oceans would be reduced, and less water would remain in the atmosphere as clouds. Large amounts of volcanism or the extensive meteor activity during the Late Heavy Bombardment (3.85 billion years ago) may have further altered the composition of the atmosphere. Remember that the forces of nature are the tools of God. As science detects the natural forces at work, we see in the biblical writings these same events described with the acknowledgment of their divine cause.

Day Three: Continent Formation and Photosynthesis (Gen 1:9-13)

Then God said, "Let the waters under the heavens be gathered together into one place, and let the dry land appear"; and it was so. (Gen 1:9 NKJV)

Science describes the bulk of the Earth's crust forming over a billion year period, midway through the development of the Earth, from three billion until two billion years ago. Remember that each day in this account represents a long period of time. The formation of continents occurred after long ages of minimal land mass.

An interesting aspect of the wording of divine commands throughout the creation week is the use of the word "let" instead of a word like "make" or "create", as if to say that it was simply allowed to happen. This does not mean that God was not the ultimate cause of these events; He commanded them to occur. It simply implies the use of natural forces in the orchestration of each creative act. As was stated earlier, the forces of nature are indistinguishable from acts of God. This is due to the fact that it is God who set up the laws of nature. And these very laws allow for Him to have control of everything, all events large and small, from the beginning and throughout time.

Then God said, "Let the Earth bring forth grass, the herb that yields seed, and the fruit tree that yields fruit according to its kind, whose seed is in itself, on the earth"; and it was so. (Gen 1:11 NKJV)

After the formation of land, God gave the command for plant life to appear on the Earth. Note that God commands the vegetation to grow from the ground. This is a strong case for a metaphorical day representing more than twenty-four hours during the creation week. (Plants take more than twenty-four hours to be fully grown.) Though plant fossils older than the oldest animal species have not yet been found, there is very strong DNA and molecular evidence for plant life by 700 million years ago, preceding the Cambrian explosion in Earth's history. Alternatively, this may have simply represented any organism capable of photosynthesis. Photosynthetic microorganisms have existed for at least 2.78 billion years as indicated in the fossil record. This would have been during the time of continent formation on day three.

Now a critic might say, "Trees did not exist until at least 400 million years ago, putting this event out of order with the other events in the creation account." The point of this passage is a generalization to indicate when plant life began. This is an account of beginnings, with plants having their beginning in the specified sequence. It was not intended to claim that all plant species currently existing came into existence at that time. The listing of a few types of plants (trees, herbs, etc.) are here to help you understand that it is plant life that is being described, since the original plant forms would have been unfamiliar to ancient peoples. Some translations use the word "vegetation" instead of "grass" and "seed-bearing plants" instead of "herbs"; the original text is not as specific as some translations imply. Even the word "seed" is used quite generally and could include spores or any other mechanism for replication. (Remember, even human offspring are described as "seed" in the Bible.) The term "vegetation" was not a rigorous scientifically defined term, so it could very well have included things like algae, lichens, fungus, and other forms of stationary multicellular life. These are known to have existed before animal life. Also, listing a few different plants makes it clear that all forms of plant life, past and present, exist due to the action of God.

Whether this passage is taken to refer to photosynthesis, primitive land plants, or currently undiscovered plant species, strong scientific parallels can be made. We cannot apply specific scientific definitions to the words in the creation story, since these texts were written long before people had any knowledge of science. Yet the creation of vegetation on day three is an accurate description of the rise of photosynthesis and plant life on Earth when compared to the discoveries of science in our age.

Day Four: Completion of the Atmosphere and View of the Sky

Then God said, "Let there be lights in the firmament of the heavens to divide the day from the night; and let them be for signs and seasons, and for days and years; and let them be for lights in the firmament of the heavens to give light on the earth"; and it was so. Then God made two great lights: the greater light to rule the day, and the lesser light to rule the night. He made the stars also. (Gen 1:14-16 NKJV)

Notice that the Sun and Moon are not named here. This is not accidental! Remember, the vantage point of this narrative is from the Earth's surface. It is the light from these bodies that is allowed to be seen clearly from the surface of the Earth now that the atmosphere was completely transparent and probably near its current composition. The Sun, Moon, and stars were created as part of the original creation of the heavens and the Earth in verse one. It is on the fourth day that the atmosphere is described as clear enough for heavenly bodies to be seen from the Earth's surface. This interpretation of day four is supported by many Bible scholars and is given as the preferred view in *Halley's Bible Handbook*.[8]

The period from 850 to 540 million years ago enjoyed a dramatic rise in oxygen levels. During this time oxygen increased steadily in the atmosphere until it reached as high as 15% by 540 million years ago. This is comparable to the present day (about 20%) and would have resulted in a transparent atmosphere. The rise in oxygen was likely caused by the proliferation of plants on land throughout this time period.

Day Five: Aquatic Life and Birds

Then God said, "Let the waters abound with an abundance of living creatures, and let the birds fly above the earth..." (Gen 1:20 NKJV)

On the fifth day God created sea life and then birds. One of the greatest mysteries in biology is known as the Cambrian Explosion. 540 million years ago the fossil record came alive. Before this time, scarcely any animal fossils have been found. And those that have, represent small strange forms of life, that have no modern analog. But at 540 million years ago we see the abrupt appearance of sea animals in great diversity. New forms continued to appear throughout the Cambrian period at the greatest rate the world has ever known. These were small and simple creatures living in the sea but represented basicly all phyla of animals living today. The Cambrian life included vast arrays of shelled creatures, soft bodied sea animals, shellfish, and vertebrate fishes. The Cambrian Explosion corresponds perfectly with the Genesis description, "Let the waters abound with an abundance of living creatures".

In the following geological periods, the oceans continue to fill with life, including large sea animals. This continues in the day five descriptions, "So God created great sea creatures..." (Gen 1:21 NKJV) Some biblical translations say "great sea monsters", a fitting portrayal of the large and horrific inhabitants of the sea during the Triassic and Jurassic periods.

Following sea life, birds come next, also on day five. Paleontologists will tell you that birds have existed since the Jurassic Period, over 150 million years ago. During this time, dinosaurs roamed the Earth. Birds are actually classified as a clade of dinosaur and are the only dinosaur group to have descendants surviving to the present. So the placement of birds on day five, is quite appropriate and rationalized by the fossil record. Critics will sometimes ask "Why is there no mention of dinosaurs in the biblical creation?" Here we see that dinosaurs are included but are described by their only remaining form, birds.

And God blessed them, saying, "be fruitful and multiply, and fill the waters in the seas, and let birds multiply on the earth." (Gen 1:22 NKJV)

Here is a command given for the created to multiply and fill the Earth, not likely in the span of twenty-four hours, but rather a very long period of time. Yet it is implied that the command is to be fulfilled on that day, just as all the other creative commands are. If these animals were meant to grow and reproduce in some hyper-accelerated fashion, wouldn't this passage mention this? Since it doesn't, the logical conclusion is that these creatures multiplied normally, through many countless generations, until their commission was fulfilled. When the creation days are viewed as ages in history, this command to multiply is logical. This can be seen as another confirmation of the day-age interpretation.

Day Six: Modern Land Animals and Humans

Then God said, "Let the earth bring forth the living creature according to its kind: cattle and creeping thing and beast of the earth, each according to its kind", and it was so. (Gen 1:24 NKJV)

On the sixth day God first created land animals. The descriptions are vague but seem to imply modern land mammals. Following the extinction of the dinosaurs, 65 million years ago, large mammals show up in the fossil record in great variety. Specifically, placental mammals that include such beasts as bears, lions, elephants, wolves, cattle, pigs, hippos, as well as creeping things like rats, mice, squirrels, all originate after the extinction of the dinosaurs.[9] Modern mammal families appeared since this time and have existed now for millions of years.

Though there are many types of creatures that may not be represented explicitly in the creation account, the use of broad categories intends to provide a general impression of the order of creation. The intent is to describe these events as simply as possible. So obviously, extinct species are not specifically mentioned. This narrative describes the most familiar aspects of life that ordinary people will recognize.

Then God said, "Let Us make man in Our image, according to Our likeness; let them have dominion over the fish of the sea, over the birds of the air, and over the cattle, over all the earth and over every creeping thing that creeps on the earth." (Gen 1:26 NKJV)

The very last event in creation is the creation of man. This is a remarkable prediction, for nearly all other creatures have existed for millions of years. But humans, the most remarkable life-form on the planet, have existed for merely thousands of years. This is not just another step in a continuously transforming world but the ultimate goal. Non-theistic scientists will say that evolution takes a random course with no goal or direction. But the Bible indicates that man is the completion of this world, the ultimate goal of creation.

The second chapter of Genesis adds more information about the events of creation, but these details are not in chronological order. Some have described the two chapters as separate creation stories. Others have claimed that they are contradictory. But the style of chapter two gives no indication of temporal order. So these passages are not contradictory but complementary.

The creation and transformation of this planet was complete. The sixth day did not end with the creation of one man but may have included many generations.

This is the book of the genealogy of Adam. In the day that God created man, He made him in the likeness of God. He created them male and female, and blessed them and called them Mankind in the day they were created. And Adam lived one hundred and thirty years, and begot a son in his own likeness, after his image, and named him Seth. (Gen 5:1-3 NKJV)

This text obviously uses the word day in way to imply a period of time in the past, and makes it synonymous with the sixth day of creation. The genealogy of Adam may also be included on this day. In the verses following, many generations of Adam's descendants are listed.

Creation Timeline

The charts on the following pages summarize the creation account and cross reference it with modern science. The time scales indicated are approximate based on the best available scientific data. The following abbrieviations are used:

Ga   gigga annum (billions of years ago)

Ma   megga annum (millions of years ago)

ka   kilo annum (thousands of years ago)

Day Zero (the first 9.5 billion years)		
Time	**Scripture Quotation**	
	Scientific Discovery	Theistic View
	In the beginning God created the heavens and the Earth. (Gen 1:1 NKJV)	
13.8 Ga	Time begins at the Big Bang. The universe expands rapidly in the first moments.	God creates time and space, and sets all the laws of physics. The universe is finely tuned for life.
13.4 Ga	Stars and galaxies coalesce from the gases produced after the big bang. (H, He)	
Ongoing	Heavy elements form within large stars. Supernova explosions spread these new elements into space. New stars continue to form and die, further enriching the universe's elemental mix.	The universe evolves by the natural laws that were appointed by God. Yet God maintains control through every phase.
4.55 Ga	A shock wave from a nearby supernova initiates the collapse of a giant cloud of gas and dust. In its wake, thousands of new stars form, including our solar system.	
	The Earth was without form and void; and darkness was on the face of the deep. And the spirit of God was hovering over the face of the waters. (Gen 1:2 NKJV)	
4.4 Ga	Earth forms and cools within about 100-200 million years. Earth has a world-wide ocean and a thick and clouded atmosphere.	God selects a water world to cultivate for life. Thick clouds shroud the Earth in darkness.

Days One and Two		
Day	Scripture Quotation	
Time	Scientific Discovery	Theistic View
Day 1	Then God said, "Let there be light"; and there was light... (Gen 1:3-5 NKJV)	
4.4-4.0 Ga	Earth's atmosphere is gradually altered as the solar wind, chemistry, cooling, and other processes transform it.	God causes the atmosphere to become translucent, allowing light to reach the surface of the Earth.
Day 2	...Thus God made the firmament, and divided the waters which were under the firmament from the waters which were above the firmament... (Gen 1:6-8 NKJV)	
3.9 Ga	The Late Heavy Meteor Bombardment. Extreme heating may have made the Earth temporarily uninhabitable as the oceans boil.	
3.8 Ga	Some atmospheric components have been lost into space. Lower temperatures return. The first microorganisms appear.	Clouds lift and water is reduced in the atmosphere.

The Third Day		
Time	**Scripture Quotation**	
	Scientific Discovery	**Theistic View**
	Then God said, "Let the waters under the heavens be gathered together into one place, and let the dry land appear..." (Gen 1:9-10 NKJV)	
3.0-2.0 Ga	For a billion years the Earth underwent massive crust building. During this period over half of Earth's land mass formed.	God creates land through plate tectonics. He controls the forces of nature.
	And God said, "Let the earth sprout vegetation, plants yielding seed, and fruit trees bearing fruit in which is their seed, each according to its kind..." (Gen 1:11-13 ESV)	
850-580 Ma	Marine plants, photo-plankton, and algae thrive on the continental shelves. Land plants may have been present by 700 million years ago. (As found by DNA evidence, though no fossils have yet been found.) This would have included moss-like species. Lichens and algae also flourished on land. (Fossils of these have been found.)	Plant life is created. Though scientists currently draw a distinction between plants, fungi, and microorganisms, ancient peoples would not. So all of these may be implied here. The point is the origin of photosynthetic life on land, not any specific species.

Day Four and Five		
Day	Scripture Quotation	
Time	Scientific Discovery	Theistic View
Day 4	Then God said, "Let there be lights in the firmament of the heavens to divide the day from the night; and let them be for signs and seasons, and for days and years...". (Gen 1:14-19 NKJV)	
700-550 Ma	Oxygen levels rise significantly as global scale photosynthesis prevails. The atmosphere becomes comparable to ours by the end of this time.	An oxygen/nitrogen atmosphere is transparent to visible light. The sun, moon, and stars can be seen from the Earth's surface. (Only the light of these appear in the sky at this time. The creation of all heavenly bodies is described in Gen 1:1, prior to the first day)
Day 5	Then God Said, "Let the waters abound with an abundance of living creatures, and let birds fly above the earth..." (Gen 1:20-23 NKJV)	
542-488 Ma	The Cambrian explosion: The massive and abrupt appearance of ancient sea animals in the fossil record.	Creation of sea animals. Nearly all animal phyla are known to have originated during this time.
199-65 Ma	Dinosaurs roamed the Earth. Theropods would be the only dinosaur group to have descendants that survive to the present. (Represented by modern birds.)	Birds have existed since the time of the dinosaurs. But only birds are listed in Genesis, since other dinosaurs did not survive to the present and would have been irrelevant to ancient people.

The Sixth Day		
Time	Scripture Quotation	
	Scientific Discovery	Theistic View
	Then God said, "Let the earth bring forth the living creature according to its kind: cattle and the creeping thing and the beast of the earth, each according to its kind..." (Gen 1:24-25 NKJV)	
65 -1 Ma	Large mammals appear in the fossil record soon after the extinction of the dinosaurs. Recent evidence confirm that placental mammals originated during this time.	God creates the familiar land mammals that still exist today. (Cattle, lions, tigers, wolves, bears, rabbits, mice, elephants, deer, etc. all originate during this period.)
	Then God said, "Let Us make man in Our image, according to Our likeness..." (Gen 1:26-31 NKJV)	
25-50 ka	Anatomically and behaviorally modern humans have only existed for less than 50 thousand years. Extremely recent in geologic time. While human-like creatures did precede us, those we would call "man" have existed only thousands of years.	Genesis lists the creation of people as the final step in creation. Not just another part of creation, but the ultimate goal.

Ancient Revelation in Scripture

Imagine that an ancient artifact, dating thousands of years old, was found exhibiting technology beyond our own. Our world view would be turned upside down. It would certainly prove intelligence from beyond Earth predating our own civilization, forcing us to reevaluate many presumptions about the world. Now something just as astounding is found within the Bible itself: predictions about the ancient Earth revealed to prophets thousands of years ago. Only in our time are these predictions testable. Divine revelation is the only explanation for this foreknowledge.

The first sentence in Genesis 1:1 states that the universe is not eternal. Since it was created, it had a beginning. This first prediction stood as a contradiction to scientific thought for much of the last century until the Big Bang theory finally prevailed over the steady state theory. With the final confirmation of the Big Bang model, the first prediction of a non-eternal universe is now verified by science.

Genesis 1:2 describes the ancient Earth long before it was a habitable place. This verse contains three predictions. The early Earth is described as a water world shrouded in darkness and devoid of continents. Though geological information is sparse for the earliest epochs of Earth's history, science has painted a very similar picture, with worldwide oceans, a thick atmosphere, and a dark surface. These three predictions are not obvious and perhaps even counterintuitive. Luck cannot explain these accurate predictions any more than luck can function as a creation mechanism. There is a reason these predictions are accurate, just as there is a reason you and I exist today.

The verses that follow portray the creation account as a series of days, not as literal twenty-four hour days, but as ages in history. This indicates the order of creative events, not the length of time over which these events occurred. Here creation is described as occurring in many steps. It describes the transformation of the Earth's atmosphere, its surface, and its biosphere, all in the correct chronological order with respect to geology and the fossil record known in our time. For these very specific events to be placed in correct order, merely by chance, would be quite impossible. Yet the order of these events was predicted more than 3000 years before science could be capable of making conclusions of its own.

Genesis creation is progressive, not an instantaneous transformation. Specifically, the atmosphere is described as changing slowly from an initial state in verse two until its completion on day four. The Earth is initially dark at its surface due to a thick atmosphere. On day one, the atmosphere thinned enough to allow substantial amounts of light through to the surface. By day two, the clouds lift so that the persistent fog would no longer reach all the way down to the ocean's surface. On day four, the sky is complete, with the atmosphere being as transparent as it is now; the stars, moon, and sun are clearly visible from the Earth's surface. Land is described in Genesis as forming on day three, originally in a lifeless state, and then transformed as it is filled with life, by progressively more advanced forms. These biblical descriptions of atmospheric and biological change are quite compatible with the geological and fossil records.

The Bible describes the waters being gathered together into one place which hints at the formation of a supercontinent, a single continent that contains all of the Earth's land mass. Due to plate tectonics, continents drift and periodically merge and break apart. Geologists believe that several times in Earth's history all of the continents were combined as a single land mass.

The last prediction is the late arrival of man. The Bible states that the creation of man was the final step in the development of the Earth. This is observed in the bones and artifacts found in the fossil record. Animals of all kinds have existed for millions of years. But anatomically modern humans have existed for less than 200,000 years, and behaviorally modern humans only 30,000 - 50,000 years.

Many primitive hominid species preceded us, but these bore more resemblance to apes in intelligence than to humans. Many animals build nests, traps, etc. Sticks and simple stone tools are used even by non-human primates of our time. Intelligence, consciousness, and creativity are features exclusive to behaviorally modern humans. The intricate artifacts produced by our ancestors expose the spark of comprehension that separates us from all other animals.

Here is a list of predictions in Genesis, all of which are now supported by modern science:

1. The universe is not eternal; it has a finite age.
2. The surface of the early Earth was dark.
3. Early in Earth's history, it was covered in a worldwide ocean.
4. The Earth lacked continents in the earliest epochs.
5. The formation of a supercontinent.
6. The recent arrival of man.
7. The formation of the Earth was not instantaneous but was progressive, each step taking time.
8. The correct order of key events in the history of the Earth are predicted:
   i. The creation of the universe, including the Sun, Earth, Moon, and stars.
   ii. The early state of the Earth's surface. (Worldwide oceans, darkness.)
   iii. The initial thinning of the atmosphere. (Light on the Earth)
   iv. The reduction of water in the atmosphere. (Separate the waters.)
   v. The formation of continents. (Dry land appears.)
   vi. The appearance of plant life on land or the development of photosynthesis.
   vii. The evolution of a transparent atmosphere. (Lights visible in the sky)
   viii. The appearance of sea creatures. (The Cambrian Explosion)
   ix. The appearance of birds. (Extant since the time of the

dinosaurs.)

x.   The appearance of modern land animals. (Placental mammals after dinosaurs.)

xi.   The appearance of modern humans.

All of these predictions were made thousands of years ago and only now are confirmed by science in our age. This result is twofold. To those who doubt the Bible, perhaps now it is worth a closer look. If the Bible's physical predictions are verified by science, there is sound reason for trusting its spiritual teachings as well. And to those who faithfully trust in the Bible's teaching, a warmer reception to the discoveries of science is warranted. There is no conflict with the broader findings of science.

These predictions make it evident that the God of the Bible is indeed the Creator we observe so clearly through cosmology. The findings of natural history agree remarkably with the Bible when a day-age approach is used for relating the time scales of the creation events. The fossil record, geology, and cosmology all confirm God is the creator of all that exists.

This correlation of science to the Bible allows us to take a deeper look into how God operates and gives us insight into some of the deeper mysteries of life. We see how God uses nature to fulfill his desire. This is not limited to creation but shows how He still interacts with the world today. He still controls the forces of nature at the most fundamental level (the quantum level). He forms each individual in their mother's womb. He governs every aspect of history, yet gives us choice, so that we possess free will and are able to have influence on the world.

Chapter Eleven
Origin of Species

Given our findings thus far, the best fit of the data shows that God created the universe and revealed in Genesis a detailed account of his role in transforming the Earth. But how does this relate to the subject of evolution? The fossil record is in excellent agreement with both the old Earth creation model and the theory of evolution. How can this apparent contradiction be reconciled? This leads to the possibility that the two may not be mutually exclusive.

Views on Evolution

It turns out that there are a number of ideas in the philosophical realm to relate scientific findings within the fossil record to theology. One of these is progressive creationism. This postulates that various species have been created progressively throughout ancient times by divine action, and rejects macroscopic evolution. Micro-evolution is accepted and accounts for breeding and some level of change in species over time. This viewpoint has the advantage of agreement with the fossil record and meshes rather harmoniously with scriptures. However, it faces some logical hurdles when considering the need to create intermediate species existent in the fossil record that are now extinct. But it could be that these extinct species were necessary for the preparation of the modern biosphere. Progressive creationism does not have a firm boundary between what level evolution is allowed and what is created through miraculous events. Also, the use of DNA from prior life-forms in the creation of new ones could explain the appearance of common descent.

Another point of view is theistic evolution. Here, evolution is just another of God's tools along with the other forces of nature. God is in control of the process and uses it to reach a specific goal. While this is logically sound and has no conflict with the large wealth of evidence for evolution and the cosmological evidence for God, it often raises objections from many who find it incompatible with traditional understandings of scripture.

Theistic evolution covers a broad range of beliefs. Many with this viewpoint see God as creating the universe with the necessary physical laws in place to produce intelligent life but rarely interacting with the natural world. These proponents will often regard the Genesis account of creation as a parable or as allegory. This viewpoint is especially common for philosophers of the science and religion interface. On the other hand, some see evolution as a tool controlled by God continually. Continuous control would be expected, based on the conclusions in chapter six due to quantum mechanics. (The uncertainty principle introduces randomness into all microscopic processes so the future is not solely determined by past conditions. Guidance along the way is required to guarantee a specific desired result.) Most importantly, the Bible describes God as being in control of all things. The analysis in the previous chapter shows a way to understand the creation passage that fits well with both theistic evolution and progressive creationism, yet it still utilizes a literal approach in rendering the creation account. The agreement of this interpretation of Genesis with science provides support for this method.

In stark contrast to progressive creationism and theistic evolution is the viewpoint of atheistic evolution. This is not to be confused with the scientific theory of evolution. Atheistic evolution is one philosophical interpretation, which is part of the atheist point of view. This contends that God has nothing to do with the evolutionary process, with evolution being the result of truly random effects, proceeding without any direction or goal. This is to say that intelligence or any other trait only results from chance and are only preserved when they provide a survival advantage. Obviously, this position would have difficulty explaining our prior findings that cosmology needs a creator to originate the universe. It also overlooks the many astounding predictions in Genesis of future scientific discovery in our time. But this is the preferred viewpoint of atheists, agnostics, and some deists. This is a common view of evolution, but is not the only interpretation of this theory.

We can see why evolution tends to be such a divisive issue, as there are so many different views on it. Young earth creationists reject it completely. Many old earth creationists accept the findings of science for the age of the earth and accept micro evolution but reject macro evolution. Supporters of progressive creationism and theistic evolution cover a spectrum of positions ranging from acceptance of micro evolution to mid-level evolution, with others supporting full macro evolution. Views on the holy scriptures of Genesis also cover a range of viewpoints, from a literal six-day account, to the days being metaphor for ages in history, to the idea that the creation narrative is simply a parable or allegory. By looking into the theory of evolution more closely, perhaps we can gain some insight into which philosophical positions carry the most merit and how to relate this result into our understanding of creation.

Darwin's Evidence for Common Descent

Let's first review the theory of evolution from a scientific perspective, separate from the various philosophical positions. Evolution describes the change of species over time through natural selection and adaptation to the environment. Natural selection follows from the simple fact that the fittest members of a population will survive and reproduce. In the event of environmental change, individuals able to best adapt will be more successful at having offspring. Small changes can accumulate over time, causing new species to eventually emerge. Modern evolutionary theory claims that all life on Earth has a universal common ancestor from which every organism has descended.

Evidence supporting the theory of evolution was plentiful in Darwin's time. There were vast varieties of species that varied across geographic boundaries. This would be most remarkable between islands. Since these islands would keep species isolated with slightly different environments, they provided a testbed for Darwin's theory. He found that each nearby island would possess similar plant and bird species, with unique adaptions for the differing conditions on each island. These species would also bear close resemblance to species on the nearest continent.[1]

Darwin also noted the remarkable success of breeders at changing the characteristics of specific breeds of animals.[2] These included dogs, horses, cattle, and virtually every animal raised for food. Plants are also commonly bred. Many varieties of flowers, vegetables, and fruits have been altered by man over the centuries for our own benefit. You don't have to splice genes to make improvements to a breed; you simply select the best of your stock for breeding. Over many generations the breed is altered. This is well documented and a widely practiced technique since ancient times.

The fossil record indicates a progression of species appearing throughout time, as expected by evolutionary theory. The most ancient life was very simple, with increasing complexity as you move forward in time. Though the fossil record is quite splotchy, there are numerous examples of intermediate forms. It is quite unlikely for the record of past life to be anywhere close to perfect. Fossilization is a chance occurrence that rarely happens. There can be vast lengths of time between fossilization events in a lineage, leaving most intermediate forms undetected.[3]

For small changes to accumulate over time to produce a new species, body parts or organs will appear or disappear gradually. If this property of evolution is true, there should be some of these intermediate parts or organs in the process of change in most organisms at any given time. These are known as vestigial parts. A popular example of this is the existence of leg bones in some species of whale. These leg bones are short, somewhat incomplete, and have no function in the modern whale. They are completely covered over with blubber, so that there is no external feature present.[4] This supports the current evolutionary thought that mammals originated on land, and marine mammals are descended from mammals that migrated to water. Another example of vestigial parts can be found in a species of blind fish that have eyes covered over by skin and are missing the lens. Dandelions, daisies, beetles, dolphins, ostriches, and countless other species have been found to have vestigial parts. Most surprisingly, even the human anatomy contains numerous vestiges of its own (the appendix, goose bumps, wisdom teeth, tail bone, etc.). Note that it is still possible that vestigial parts have some current use for the present form, but the original usage has been lost.[5]

Basic Principles of Evolution

Though the DNA molecule was not known in Darwin's time, the transfer of traits to offspring has been understood for thousands of years. What distinguishes evolution is that traits can change over time within a lineage. These random mutations may be small or severe and are what makes evolution possible. Without a natural way to vary features or traits, there would be no mechanism to allow organisms to evolve.

The basic driver of natural selection is competition. Organisms compete for limited resources in their struggle to survive. Sometimes this struggle is due to a limited amount of food that is shared by competing individuals or species. Other times the competition is avoidance of predators. The environment in which an organism lives creates unique challenges to survival. These challenges drive the selection of characteristics that give the organism an advantage due to improved chances of survival and reproduction. Environmental changes can then lead to new selective pressures that drive change in a species.

All species reproduce at an exponential rate, that is, with more offspring than parents.[6] This would lead to rapid over population if it were not for premature death due to competition and predation. When a species finds itself to have great advantage over other species in its environment, it will increase in numbers exponentially until its members are consuming all the resources available to it. At this point, it comes into equilibrium with the environment, when not all individuals will be able to find enough food, shelter, or some other limiting resource.

Natural selection is the tendency of nature to select the fittest individuals for survival and reproduction, thereby selectively choosing the most desired traits. If a population is in equilibrium with its environment, competition is tight. When the population was growing, differences may accumulate due to excess resources; even the weakest will still find food. But as competition increases with scarcity, natural selection begins to take place. The individuals that are best suited with the most advantageous traits survive, the others do not.

Some minor mutations may be beneficial to survival. If they are, they may come to be required for survival due to competition and natural selection. These principles drive species to change. Populations of the same species that are separate from each other for a great deal of time can eventually look quite different due to different traits that have developed and different environmental conditions.

This occurs with plants and animals that are bred and separated from those in the wild. The results of natural selection can be compared to that of breeding, except natural selection acts on behalf of the organism, while breeding acts to meet the goals of the breeder.

These micro-evolutionary effects can be extrapolated to the whole of life on Earth. This is the basic claim of the theory of evolution – that all life on Earth descended from one or just a few progenitors.

DNA and Evolution

We have already listed many compelling pieces of evidence in support of evolution. But with the knowledge of DNA and the ability to map an organism's entire genome, new insights abound. We see that all life on Earth use a common genetic structure and the same 20 amino acids that are shared universally.[7]

Deoxyribonucleic acid (DNA) is the molecule of life. It is a data storage device that contains all of the information to construct and regulate the function of every part of an organism. These molecules are paired to form a double helix, like a ladder that has been twisted into a spiral. Each rung of the ladder represents a base pair, two bases connected at the middle, each representing a "letter" in the sequence of the code. The four bases which make up the genetic code are: Adenine (A), Thymine (T), Guanine (G), and Cytosine (C). These can be linked in any order down one side of the ladder. The other side of the ladder must join A to T, and T to A, and likewise G to C, and C to G, so that A always pairs with T, and G always pairs with C.[8]

A chromosome is a pair of DNA molecules wound together and joined loosely at the rungs forming the double helix. A typical DNA molecule is millions of nucleotides (bases) long with every three bases forming a codon or "word" in the sequence. Having three-letter words and four letters to choose from, there are sixty-four codons available for use in this coding language.[9]

A gene is a specific region within the DNA molecule that contains the blueprint for constructing a specific protein. Proteins are the building blocks of life. As much as half of the dry weight of a cell can consist of proteins. Proteins can be structural, as found in the cell wall, chemical, when used as enzymes that catalyze chemical reactions, mechanical, as in myosin that produce force in muscle fibers, and also many other functions.[10] A typical gene will consist of a sequence of several thousand codons which indicate the order of construction for a protein out of amino acids. There are twenty amino acids which are used to build proteins. Several amino acids have more than one codon to code them, as there are more codons than there are amino acids. There is one codon that is used as a start command, indicating the start of a gene, and there are three different stop codons.[11] As you read the codons of a gene, the order for chaining amino acids together is given for forming the protein. Which amino acids are used and the order that they are strung together determines the protein's final properties. With all the organization of a computer program, the DNA acts as a hard drive.

To create a protein, the chromosome is unwound so that the rungs are broken apart in the area that contains the gene. The gene is then transcribed onto a single stranded molecule called mRNA which creates a copy of the build sequence. This molecule is joined with a ribosome, tRNA molecules, amino acids, enzymes, and an energy source to drive construction. When the entire sequence has finished, a new protein is created and will find its place in the cell. The tRNA return to the cell fluid (cytoplasm) to again acquire an amino acid and then be available for building another protein.[12] This entire process is described in more detail in the next chapter.

When the DNA sequences of a chromosome are read, hundreds of genes may be found, but that is not all. The genes are initiated by start codons and terminated by stop codons, but after the stops and before the starts, there are vast series of bases that do not code for proteins. When the human genome was first sequenced, surprisingly few genes were found. In fact, more than 97% of the human genome does not code for proteins.[13] At first, researchers thought that the vast majority of the genetic code was simply "junk DNA". That is, left over sequences of genes long ago mutated that no longer had any current function. From 2003 to 2012 the ENCODE project, including thirty-two institutions, endeavored to determine if these extra DNA sequences did or did not have any current use. To the surprise of many, they found that at least 80% of the human genome performed some necessary function. Some DNA regions act to regulate the use of genes. Others code for RNA sequences that do not become proteins. This was the real surprise. A full 76% of human DNA is transcribed into RNA, though only a small percentage is then translated into proteins. These various RNA molecules have a variety of functions, some regulating the transcription of RNA, the construction of proteins, or the control of very specific functions in specific parts of the cell.[14] We are only just beginning to understand how the program encoded in our DNA is executed in the control and construction of the body's many cell types.

Certain genes known as "jumping genes" have control sequences that can cause the gene to occasionally be cut and pasted, or copy and pasted into another part of the genome.[15] Often the duplication of the gene has no detrimental effect. Other times, it can split or overwrite a useful gene or control sequence. These genes can be considered selfish DNA parasites as they can clutter the genome with copies of themselves. Sometimes the jumping gene is beneficial by preserving a copy unchanged when a mutation occurs in the original gene. When these duplicated genes are copied and subsequently mutated or truncated, they become ancient repetitive elements (AREs), that is, non-functional, left over damaged duplicates of the original gene.[16] These sequences are passed on to offspring in their DNA. These broken pieces of the jumping genes can mutate and may in subsequent generations have some use in control or as an entirely new gene. It is believed that nearly half of the human genome is made up of jumping genes and AREs.[17]

Another component of an organism's genome is the pseudo-gene. This is a non-coding region that was once a functional gene, but due to mutation, it no longer works. These pseudogenes litter DNA and serve as a record of past genes. Closely related animals will have most genes in common, with missing genes usually repre-sented by a modified functional gene or a non-functional pseudo-gene. The more distantly related two animals are, the more dissimi-lar the genes and pseudogenes will be.

When comparing genomes of different species, many genes will be in common. For example, humans and chimpanzees have 96% of their DNA in common.[18] The similarity does not stop with primates, but many genes and the order they occur remain shared with all other mammals, and some genetic code is remarkably similar in all forms of life. Not only are DNA sequences similar, but they show gradual morphology from one creature to another. The closer related the species are, the more closely the genes and AREs match up. Dis-tantly related creatures have very different genes and DNA structure, but certain commonalities remain. Closely-related species are almost identical in chromosome count, gene order on the chromosomes, and positions of pseudogenes and AREs. This provides yet another line of support for common descent.

Because DNA that does not code for proteins is less critical, it is more tolerant of change without negative effect. This provides a record of past mutations that geneticists can read like a book. By comparing genomes of different species, specific mutations can be found that led to their divergence from a common ancestor. By com-paring pseudogenes and AREs in various species, the order of each mutation can be found, providing an indication of the order that each species diverged from their last common ancestor. DNA is copied with remarkable precision, with only about one error for every 100 million bases per generation.[19] This provides a way of approximating the time since two lineages diverged from a common ancestor. In addition to storing all the information needed for building and maintaining an organism, the DNA also stores a detailed account of the organism's ancestry.

Accelerated Evolution?

Here on Earth, evolution progresses rather slowly. But China has been pursuing a breeding program which involves sending seeds into space to be exposed to microgravity and radiation. This exposure can increase the natural mutation rate hundreds of times. The mutations are often detrimental, but in a few of the seeds, a beneficial change is induced. After return to Earth, the seeds would be grown, and the desired traits selected for continued breeding. It takes several years to cull out the undesirable traits and capture the beneficial ones. But this is ten times faster than traditional breeding. The results of this program are astounding, altering the size, flavor, vitamin content, and yield of the new breeds. This includes tomatoes that are as sweet as oranges, eggplants the size of basketballs, and cucumbers a half meter long. There are peppers that are so improved that their use has spread across many provinces. China hopes to improve the food supply through these mutant vegetables.

Problems with Evolution

Though it is often suppressed in scientific circles, there are problems associated with the theory of evolution. These may not be death blows to the theory but are significant enough as to require some caution before ruling out competing ideas. Darwin himself recognized some of these issues and pointed them out honestly.

One glaring concern to Darwin, was the Cambrian explosion. This is still a challenge to evolutionary theory to this day. This event was known during his time and is quite obvious in the layered strata.

The fossil record does not show a few scattered fossils deep in the layers, increasing gradually until fossils become common, as you might expect from evolution. Instead, when looking at a cut away section of layered deposits, fossils are found all the way down to the Cambrian, where there exists an abrupt boundary between rocks with animal fossils and rocks without.[20] This shelly layer contains all major existing and extinct animal phyla.[21] (The animal kingdom is divided into twenty-six or so animal phyla.) While there are many different ideas to explain the rapid advance of life during the Cambrian, it still represents a remarkable event that challenges standard evolutionary thought.

What is also notable about this event is the lack of intermediate forms preceding these first animals and also for subsequent life-forms appearing later. This is often an over-used argument against evolution, but it still requires an explanation. The fossil record is much more stepwise than you would expect from the gradual forces of evolution. One common explanation for this is that fossilization is a chance occurrence. So we do not see every point along the way but only get to peek in at random points in history.[22] Another explanation for this is that environmental change drives species to adapt quickly or else they become extinct. Both of these explanations are plausible reasons, but neither satisfactorily explains this event nor completely nullifies this issue.

DNA has offered many lines of evidence in support of evolution. However, a significant problem arises when you map out various genes between the three major domains of life. It seems nature has played a trick on us. We expect that genes are passed from parent to child all the way back to the last universal common ancestor. Changes in genes that occur in one domain should be passed to children but not to cousins in the other domains. This is not what we find. Genes that originated in bacteria are found in eukaryotes. All sorts of genes are found out of place, not originating in their ancestors.[23] This makes constructing an evolutionary tree of life difficult for the early stages. The assortment of genes might look more like competing car designs, where the manufacturers are copying desirable features from each other, rather than resulting from common descent.

This is where the idea of lateral gene transfer comes in. To explain these out of place genes, biologists propose that early on the various domains of life exchanged genes between each other. This is demonstrated in our time by bacteria, who can pass antibiotic resistance to other species of bacteria through gene swapping.[24] This can happen when a bacterium takes in DNA found in its environment and incorporates the DNA into its own genome. Also, DNA can be transferred by viral infection. It is proposed that early on, all three domains of life swapped genes in this way. Lateral gene transfer offers a possible explanation for these out of place genes, but this could also be explained by arguments from intelligent design.

Other insights from DNA and cellular function show that there is remarkable intelligence in its design. DNA reads like a computer program, and the processes within the cell that utilize this program have more complexity than a super computer. Yet all life requires this complexity as a minimum threshold for sustaining and replicating itself. This leads us to wonder how life got started in the first place. This will be covered in the next chapter, as this subject is outside the scope of the theory of evolution, which only deals with the change in life-forms over time.

All of these challenges to evolution are real, though they may not outweigh the extensive volumes of evidence in support for the theory. But these challenges are difficult to account for with an atheistic stance on evolution. If God was involved in guiding evolution or if God progressively created new life-forms with borrowed DNA, then we would expect to see these irregularities in the fossil record and in the genomes of all life. So what appears as problems for evolution are really only problems for the atheistic philosophy. This lends support for theistic explanations of origin, especially various forms of theistic evolution or progressive creationism.

Spiritual Implications

The scientific theory of evolution does not support or refute atheism. It does not even address the origin of life, only the evolution of life after it is established. Though evolution is popularly used as a reason against God's existence, this claim is due to a misunderstanding of the creation account. There are a number of reasons why there is no atheistic support from the study of evolution. Many of the findings of evolution actually support God's role and involvement in the creation of life.

Evolution could be implied in the creation account by the phrase, "Let the Earth bring forth", used often throughout the passage. The use of the word, "Let", suggests a command to the forces of nature to produce the desired effect. This in no way diminishes God's role as Creator, since all of natural history proceeds according to His plan.

The same evidence found in the fossil record that supports evolution also matches up with the order of creation as found in the first chapter of the Bible. This prediction of future scientific discovery proves its divine authorship. Evolution is quite compatible with the written word of God in this respect.

We have already seen how the forces of nature are the tools of God, being fine-tuned for life from the foundation of the universe. Evolution could be another of these tools. Mutations occur at the molecular level, where quantum fluctuations yield their effect. Since these quantum fluctuations can be utilized by God as an invisible control mechanism, He maintains dominion over the whole process. While the theory of evolution predicts only a random continuum of change, it does not include the spiritual side of the equation, which is the guiding influence of God.

For evolution to occur, there are a number of prerequisites. First you must have life; evolution does not address how the first life-form came to be. Second, that life must be able to reproduce copies of itself. This is no small order. For it also has to replicate the ability of self-replication. Even with all of our advances in technology, we are unable to make even a simple machine that is able to replicate itself in this way. Third, there must be a built-in mechanism allowing for variation. This cannot be too slow, or evolution could not make substantial progress in the lifetime of a planet. Slow variation also would not be able to keep up with the pace of environmental change that occurs throughout a planet's life cycle. Too rapid an evolutionary rate would not provide the stability necessary to ensure that offspring survive to reproduce. Most mutations are harmful, so a rapid rate of evolution would lead to the rapid decline of life. The evolutionary rate must be just right for progressive beneficial change to occur.

All life on Earth is based on the same code represented in DNA. The same basic procedure to exploit DNA for self construction and reproduction is also used by all known life. This suggests that it may be the only natural mechanism possible for life. If any of the characteristics of DNA, amino acids, solvents, and atoms were different, life might not be possible. The simplest specification for life we know is extraordinarily complex. A bacterium requires thousands of genes[25] and includes all the complex machinery for self replication. There is very little that could be pared down from the bacterial structure without jeopardizing its ability to survive and reproduce. There is no satisfactory natural explanation for the origin of DNA or the many cellular structures that process its instructions. The fitfulness of DNA for life is only possible due to the many properties of matter that have been divinely appointed for the purpose of constructing life-forms. The ability to change over time is built into the DNA molecule. Life is *designed to evolve*.

My friend supporting a young earth viewpoint has asked, "Why would God need evolution?" But the same question could be asked of a carpenter about his preference of a power saw over a hand saw. His answer is, "It's faster, easier, and makes a cleaner cut." In the same way, it may be that God's use of the forces of nature are to accomplish a certain desired result. A different method of creation would have had a different result. If in no other way, at least the fossil and DNA records would reveal this alternate history.

In light of the coincidence of life and discovery in the universe, evolution makes an excellent way to track creation so that man could discover how God created life on Earth. In our discovery of creation, we learn about the nature of God. We can see that He is long suffering and patient in working through evolutionary time scales before creation is complete. Scientific discovery allows for the order of creative events to be determined, that correlate to the Genesis account. By this, confirmation of the Bible via science offers further revelation of God. So through evolution and other sciences as well, God makes His glory known to His creation.

My atheist friend would say, "Why does evolution need God?" It is difficult to explain the progression of life and the emergence of intelligence without God. The irregularities within the fossil record and the masterful complexity of the genetic code are best explained with God as designer. Someone who cannot accept God will have a different view of things because they have filtered out the spiritual side. But they are left with big questions as to how such miraculous results occur purely by chance.

The current debates between evolution and creationism are fueled by our own prejudices. In the narrative that follows, one man denies the claims of the artist in residence, while another takes his word too literally.

Leonardo DaVinci lays down his brush, sits down, and gazes with satisfaction at his new creation. Many years of toil and late nights had yielded its reward. He thinks to himself, "It is very good. This is the greatest work I have ever created." Just then a man walks in the room, his jaw drops as he beholds the Mona Lisa. For some time he just stares at it, analyzing its every stroke. Finally he proclaims, "Behold the wonder of this paint and the creativity of this brush!"

At this DaVinci is quite insulted but yet humbly replies, "Good sir, I do beg your pardon, but this masterpiece is my creation, the work of my very own hands." But the man immediately declares, "You scoundrel, you pretender! I can plainly see the colors of the paint on this pallet. They match perfectly the colors that are on this painting. Behold, the brush is still wet with the paint of her face. The strokes on this canvas are a perfect match to the bristles of this brush. Nowhere are the fingerprints of a man found on this canvas, not a single smudge. Certainly this masterpiece was born of this paint and this brush and was not the work of your hands!"

Now another man, a friend of DaVinci was outside the door and overheard the whole conversation. He entered the room and said, "I know this man DaVinci, that his word is true. Therefore he certainly painted this whole painting with only the fingers of his hand, and did not even use a brush or paint. For he said it was the work of his hands."

As ridiculous as this story sounds, this is just what we have with the current debates on evolution. Some deny obvious evidence for the Creator of the universe and then credit our existence with the tools He used. Others faithfully trust in his word but take some statements more literally than was intended, thereby missing out on the discoveries of how God works in our world and throughout the universe.

Supporters of Theistic Evolution

A Gallup Poll on evolution was taken in 2010 as it has been done in past years. This poll asked people what they believe about the origin of man. Here, 40% of Americans said that God created humans in their present form. 38% of Americans believe that humans evolved but God guided the process, in other words, theistic evolution. And 16% of Americans claimed that humans evolved and that God had no part in it; this is atheistic evolution. The remaining 6% had no opinion or some other belief. Results from this same question in 1982 showed the same percentage for evolution with God's guidance. Atheistic evolution has risen from 9%, while those not believing in evolution at all has declined from 44%.[26]

This shows that the majority of Americans still believe in God (78%) and that a smaller majority believe in evolution (54%). While popularity does not dictate truth, it does indicate that theistic evolution is not simply an extremist position, grasping at straws to make science and religion compatible. Notable theologians such as C. S. Lewis and Pope John Paul II were supporters of theistic evolution. Even the famous evangelical preacher Billy Graham accepted theistic evolution as a possible interpretation of scripture. It is also supported by many Christian scientists, such as Francis Collins, director of the human genome project. He actually came to a belief in God through his study of evolution.[27] He could not see any other way for it all to work out so perfectly, except through direction by God. Other prominent researchers and biologists from all over the world support theistic evolution. This is not to say that atheism is not common among biologists; it is. But there are also thousands of Christian scientists who find harmony with the Bible, science, and evolution.

For creationists that reject evolution on theological grounds, there are scientifically plausible alternatives to evolution. Progressive creationism aligns with the age of the earth and universe and the fossil record. Though most fossil and DNA evidence currently supports evolution, future discovery could one day tip the scale for the special creation of individual species or groups by God, as in progressive creationism.

As was shown here, evolution need not be in conflict with the

Bible, nor should it impact theology. The creation account need not be taken as parable. It is compatible with evolution even when taken as a literal historical account.

The belief that the Bible contradicts science, especially as it pertains to evolution, is popularized by the media. Conflict is more interesting than harmony. But as witnessed by the polls, many accept both God and evolution. And it is not only bystanders who take this position, but also those who are among religious leadership and leading scientists alike.

Evolution or Revolution?

Though many are content to find harmony between evolution and God, others object to such a pairing. To mention evolution often instills contempt from some in the religious community. At the same time, adding theism to the theory of evolution might cause many evolutionists to cringe. But scientific justification for God is strong as ever, and neither can the evidence for evolution be ignored. Some degree of evolution must be accepted, even if only at the micro evolutionary level. Most forms of old earth creationism accept the adaptation of life through micro evolution. Progressive creationism may allow for common descent by divine modification or creation using DNA from prior life-forms, though still holding out against full macro evolution.

As it is, there are scientific reasons to explore both progressive creationism and theistic evolution. The biblical creation account lines up with these views as well. What can be said is that the atheistic form of evolution is in trouble. Unguided, purposeless evolution has difficulty explaining the many anomalies in the theory. Evolution does a very good job of explaining the adaptation of a species to its environment. This allows for life to survive environmental change. Evolution effectively explains genetic patterns and similarities between various life-forms. It also explains the success of breeders over the centuries at developing domestic livestock. These breeds are very different from their wild progenitors and are greatly enhanced for human use.

But when we look at major events and abrupt changes in the fossil record, evolution runs into difficulty. The Cambrian explosion is just one example where the situation looks more like revolution rather than evolution. Perhaps God introduced revolutionary new life-forms in an abrupt fashion at various times to support the Earth's development? Alternatively, God may have guided evolution down a path that lead to the rapid appearance of animal phyla during the Cambrian. The difference between various philosophical perspectives may not be testable scientifically, since science only deals with the physical aspect of reality. However, unguided evolution would have great difficulty accomplishing the many revolutionary events that led to advanced life and the emergence of intelligence on Earth.

Another major problem for atheistic evolution is similar to the problem faced by atheistic interpretations of the Big Bang. How did it all start? Like the evolution of the universe, the evolution of life had to have a beginning. Evolution does not work on non-living material. Evolution needs life to exist before it can do its job. The origin of life is a great mystery in science and is a separate subject of its own. Once life began, a biological Big Bang may have came forth from simple beginnings. And like the cosmological Big Bang, theological implications abound.

Chapter Twelve
Origin of Life

The origin of life is still one of the greatest unsolved mysteries of science. This question goes far beyond the scope of evolution, since evolution cannot take place until life is already established. What we do know is that life existed on this planet soon after its formation. Too soon, according to many biologists, for life to have resulted from purely terrestrial mechanisms. Even if ample time was available, it is still unknown what process ultimately led to life on Earth.

Abiogenesis is the formation of living biological material from purely non-living sources. This process has never been observed in nature. Abiogenesis is the prime goal of all origin of life research, to discover how ancient lifeless material became living.

This area of study becomes particularly interesting as any findings could be extrapolated to the likelihood of finding life elsewhere in the cosmos. So anything we learn about life's origin on Earth gives us information about life's potential beyond our terrestrial home.

All Earth life is based on DNA, RNA, and the same twenty

amino acids.[1] Without these nucleic acids, life is not possible. Though some have suggested that there may be other chemistries of life, there are no indications or evidence for this found in nature. Lacking any workable theories or evidence for alternate biologies, we are faced with the fact that there is, most likely, only one chemistry of life. This is why searches for extraterrestrial life focus on finding conditions favorable to carbon-based life-forms and Earth-like conditions.

The Chemistry of Life

All life on Earth use the same four bases as letters in three letter words to form codons. The resulting sixty-four possible combinations code for the same twenty amino acids used by all Earth life. The chart to the right shows which DNA codons correspond for each amino acid or stop command.

The simplest organisms capable of self-sustained reproduction are single-celled prokaryotes such as bacteria and archaea.

Our Genetic Code

TTT TTC — Phe	TCT TCC TCA TCG — Ser	TAT TAC — Tyr TAA TAG — Stop	TGT TGC — Cys TGA — Stop TGG — Trp
CTT CTC CTA CTG — Leu	CCT CCC CCA CCG — Pro	CAT CAC — His CAA CAG — Gln	CGT CGC CGA CGG — Arg
ATT ATC ATA — Ile ATG — Met	ACT ACC ACA ACG — Thr	AAT AAC — Asn AAA AAG — Lys	AGT AGC — Ser AGA AGG — Arg
GTT GTC GTA GTG — Val	GCT GCC GCA GCG — Ala	GAT GAC — Asp GAA GAG — Glu	GGT GGC GGA GGG — Gly

*The horizontal text lists all possible 3 letter codons in our genetic code. The vertical text indicates the corresponding amino acid or a "stop" command. Met is an amino acid that acts as the "start" codon.

Their simplicity is only relative to the much larger cells of eukaryotes, such as in plants and animals, as even the simplest cells are still highly complex. The cell membrane must allow essential nutrients and food to pass through, and it must keep contained any materials it manufactures for its own use. It also must keep unwanted substances out. Within these cells is a wonderful brew of necessary enzymes, miniature structures, and chemicals. Some of these chemicals supply energy that must be extracted from food to power the cell's many chemical processes. All of this is in addition to the DNA, RNA, and various other components required for their construction.

It is difficult to grasp the complexity of even the simplest of life due to its complicated chemistry and microscopic size. But when you consider its prime function, which is to reproduce, it's easy to see just how complex it is. For with all of our current technology, we are (presently) unable to make a machine that can reproduce both itself and its reproductive ability. In addition, the first life-form would need to have the ability to acquire its own raw materials and utilize an energy source to grow and reproduce.

RNA (ribose nucleic acid) is a complex chaining molecule necessary for the transfer of genetic information within a cell. It is the property of stringing together long chains of molecular groupings, called nucleotides, that allows it to store vast amounts of genetic data. RNA nucleotides consist of the sugar ribose, a phosphate group, and one of four bases. RNA acts as a working blueprint of the information in DNA, since DNA information is not directly accessible to the rest of the cell. The four bases for RNA are adenine (A), guanine (G), cytosine (C), and uracil (U). Uracil replaces thymine (T) from DNA.[2] The other bases are also used in DNA. The RNA nucleotides are strung together in long chains with each base representing a different letter in the genetic code. Three of these letters form a word (codon) which corresponds to the codons in DNA.[3] The following chart depicts the process of how the cell copies information stored in the DNA and creates RNA chains.

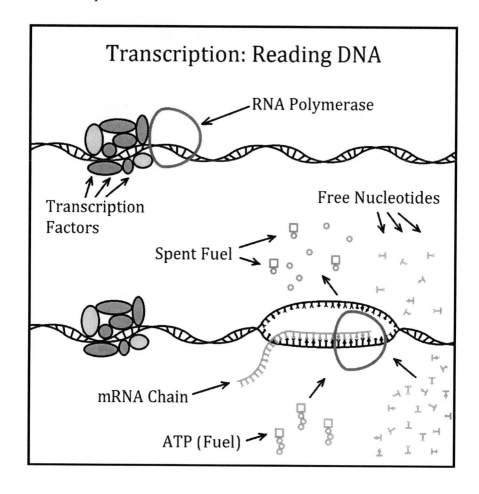

Transcription: Reading DNA

RNA Polymerase

Transcription Factors

Free Nucleotides

Spent Fuel

mRNA Chain

ATP (Fuel)

Messenger RNA (mRNA) are created by copying genes from DNA for the purpose of constructing proteins; this is called transcription and is common to all Earth life. This is done when enzymes selectively unzip a portion of DNA. Transcription is regulated in the cell by several transcription factors, proteins that must bind at an initiation site preceding the "start" codon for the gene. An enzyme called RNA polymerase is guided to the starting region by these transcription factors where it binds to the DNA, unzips it, and initiates the copying process. The polymerase moves along the DNA, unzipping the DNA and adding nucleotides to the mRNA as it moves. Adenosine triphosphate (ATP) is the primary energy transport molecule of the cell. ATP molecules are broken to supply the energy required as each nucleotide is added. After the polymerase has past a region, the DNA zips back up behind it. A specific series of bases is used to mark the termination site on the DNA. When the termination site is reached, transcription stops, and the mRNA is released into the cell.[4]

While some RNA molecules are used directly, without further processing, others are slated for protein construction. Proteins are the building blocks of life; as much as half of the dry weight of a cell consists of them.

The mRNA molecule will consist of thousands of nucleotides, three nucleotides (bases) per codon. Each codon in the sequence codes for one of twenty amino acids. Transfer RNA (tRNA) is another essential device consisting of short RNA strands about eighty nucleotides long. Of this, only three nucleotides, representing the mating sequence for a single codon, are free to attach to other RNA molecules. There is a unique tRNA molecule for each codon in the genetic code, except for the stop codons. There are no tRNA molecules for these. Each tRNA is joined with its appropriate amino acid in a process requiring a unique enzyme specialized for that particular tRNA and powered with energy supplied by an ATP molecule.

When an mRNA is produced for protein synthesis, the cell's construction machinery is put to work. The tRNA molecules are recycled after use, since they become stripped of their amino acid when it is added to the new protein. Many enzymes and other proteins are required in the translation process. Energy is also needed, supplied by a guanosine triphosphate (GTP) molecule.[6] The following chart provides a simplified pictorial of how proteins are constructed within the cell. This shows the various stages of manufacturing a protein.

Translation of Protiens

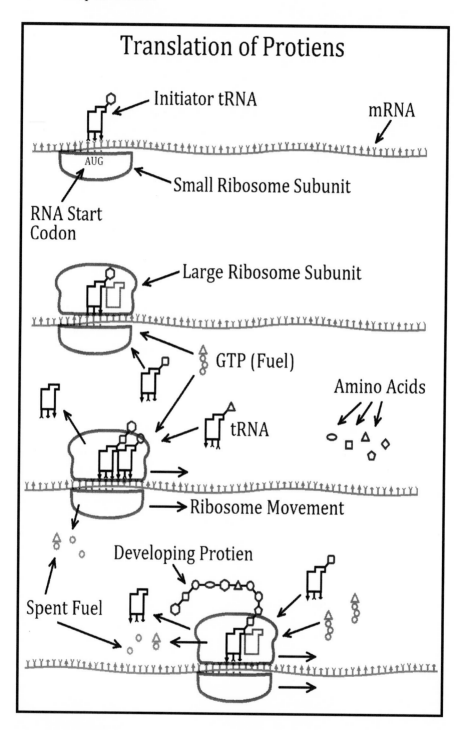

A small membraneless organelle called a ribosome is responsible for translating the information on the messenger RNA (mRNA) and constructing a protein.[7] The ribosome consists of two parts, each containing many proteins and RNA molecules. The smaller part acts as an initiator that attaches to the initiation site on the mRNA strand. A tRNA binds to the start codon, then the larger ribosome subunit attaches to complete the ribosome. The ribosome effectively reads the mRNA and catalyzes the attachment of amino acids, moving along the strand as it works. The correct tRNA mates up to each codon in the mRNA sequence and adds its amino acid to the growing chain. This builds up a massive molecular chain known as a protein. It takes only sixty milliseconds to add each amino acid.[8] That would add up to about a minute for a protein consisting of 1000 amino acids. Once the entire build sequence has completed, the protein detaches and folds up in a way specific to its own design. A protein is a single molecule that may contain millions of atoms. There are thousands of different proteins in an organism, and each organism has its own specific build plans for its own proteins. Proteins are the basic building blocks of cells; they also act as chemicals for all sorts of processes. Some proteins are the very enzymes needed for processing DNA and RNA and those needed for transcription and translation during protein construction.

All of the processes described above require energy. When tRNA is linked to its appropriate amino acid, an enzyme is required to bring the tRNA, the amino acid, and ATP (the energy source) together. The ATP molecule is broken, releasing energy that is used to link the tRNA to the amino acid.[9] Similarly, many of the other enzymes, as well as the ribosomes, require specific energy-supplying molecules in addition to the component parts of the reactions they facilitate. A steady supply of energy, an effective means for storing it, and mechanisms for transporting it are essential to all of life's vital functions.

The Search for Abiogenesis

All biological enzymes are proteins.[10] This leaves us with the proverbial chicken and the egg problem. Enzymes are needed to synthesize DNA, RNA, and proteins, but these enzymes cannot be created without the DNA, RNA, and the enzymes themselves already available. The complexity of these reactions and all of the components required for self-replication make the chance occurrence of a self-replicating system an exceedingly improbable event. There are literally thousands of different substances that are required by the simplest cells to live and reproduce. These substances are manufactured by the cell itself, and each one is intricately complex. A protein is so complex that it is rather inconceivable that one could be accidentally made by non-biological processes. So much more so is the cell itself, with millions of proteins and nucleic acids working together as a living entity. This is what makes the natural origin of life so hard to comprehend. The fact that there is only one line of descent on Earth and, therefore, only one incarnation of life known, shows how rare this event was.

Some biologists (especially oceanographers) speculate that life may have began in hydrothermal vents miles below the surface on the ocean floor.[11] These deep sea vents can spew toxic chemicals and scalding hot water into the dark abyss. In spite of this inhospitable environment, an ecosystem of its own exists. Here lives many strange forms of microbial life called extremophiles. Though these odd organisms live in isolation from the rest of the world, they also share common descent with the rest of life on Earth. This has led some to speculate that it is this sort of place where life may have begun.

Other biologists have reasoned that life may have begun in ice.[12] This is due to the tendency of many organic molecules to break down over time, thus limiting their ability to accumulate in the pre-biotic environment. In colder conditions, they argue, these molecules would last longer and have better chance of randomly producing the first self-replicating entity. In addition, when liquids freeze, the materials dissolved in them freeze last and are therefore concentrated by the freezing process. Often, pockets of liquid water remain within the ice, containing a concentration of impurities. These chambers could act as tiny test tubes where all sorts of chemical experiments could play out. So this provides a proposed route to life from non-living material.

Panspermia is another popular idea. In recognition of the insurmountable odds of life forming by chance on Earth, it is reasoned that life may have been transported here from space. This would have happened through meteor or comet impacts bringing microorganisms from a distant world. The odds for abiogenesis improve when you include many billions of planets and the entire history of the universe for the random creation of life. Backing for this idea comes from several experiments that have been conducted to determine the survival rate of various microorganisms. Indeed, many microbial life-forms have been found to be quite hearty in space. Of course they remain dormant without water or air. After a few weeks in space, the survivors can be rejuvenated upon return to Earth by receiving the nutrients, water, and air they require. The problem with this idea is that transport between solar systems would take millions of years for material exchange to occur. In this amount of time, any frozen microbes would certainly be dead due to break down of their DNA, proteins, and other organic molecules essential for survival.

Some have suggested that life may have originated on Mars and was transported here through meteor impacts. Though it would be very unlikely for a microbe to survive ejection by a massive impact and subsequent falling to the Earth a few years later, plausible mechanisms have been proposed. But this would not help out the incredible odds against life's first fusion, since Mars formed at the same time as the Earth. Given that Earth has had an extensive biosphere for the last 3.8 billion years, it is more likely that life from Earth has infected Mars than the other way around.

Some biologists favor the theory that life began as RNA based rather than DNA based. In this scenario, RNA alone acted as the repository for genetic information and as its own catalyst for replication. There have been many successes in generating some biochemicals in the lab from purely non-biological materials. Also some of these and others have been found in space. These include many amino acids and the bases adenine and guanine.[13] Getting RNA to synthesize initially could be challenging. But once you have an RNA molecule, it has been suggested that it could naturally make copies of itself. This has been demonstrated in the laboratory by showing RNA pairing with free nucleotides in solution. But this only represents half the process, since pairing produces an RNA with an opposite coding of the original. To get the original, this opposite would need to be separated and re-paired with free nucleotides to complete the copy process. This has not been done successfully in the lab after years of trying, without the help of protein-based enzymes.[14]

Another challenge exists for life's first synthesis. Many organic molecules can be made in right or left-handed versions. These experiments have been done with all right-handed nucleotides as used by all Earth life. Non-biologically produced nucleotides would be expected to have an equal mix of handedness. But mixing both types into solution inhibits the copying process.[15] In addition, the nucleotides themselves would be difficult to construct without enzymes. An energy source in the correct form is also required. Even in ideal circumstances created in the laboratory, we are still not able to duplicate the sort of reactions that would have been required for the first self-copying RNA.

There are so many challenges for the first origin of life on Earth that it becomes clear that we really do not know, scientifically, how life began. The late origin of life researcher, Leslie Orgel, has stated this opinion: "The full details of how the RNA world, and life, emerged may not be revealed in the near future."[16] This is not to say that we will never discover how it occurred, but it is a statement of the current level of scientific ignorance on the subject.

The Creator's Imprint on Life

Life has existed on Earth for at least 3.8 billion years.[17] If the late heavy bombardment of 3.9 billion years ago produced intolerable conditions for life at that time, then life seems to have appeared as soon as the Earth became habitable. We do not have fossils from this early time, just chemical signatures indicative of life. The earliest fossils preserved to the present day date back to 3.6 billion years ago and look identical to bacteria existing on Earth today.[18]

It is truly fascinating that life in all its complexity could arise so quickly. Equally amazing is that it did this only once.[19] If the rise of life was an easy process, as some biologists have suggested, then why is all life descended from a single common ancestor? If it was easy, one would expect that it would have occurred many times, and these separate lines of descent would be apparent today. But this is not the case; all organisms share a common mark on their genomes, as witness to their brotherhood. Some have suggested that it may have arisen more than once, but only one strain of life survives to the present. This is unlikely given the resilience of life and the vast array of Earth ecosystems. Somewhere a strain of these past forms or their fossils would remain if, indeed, they ever existed.

This results in a paradox from a purely natural perspective. The first incarnation of life must have been an unlikely occurrence, or else there should be many separate lines of descent. Yet it happened on Earth almost immediately after ocean vaporizing impacts had subsided. As soon as life could survive, it was born. We are extraordinarily lucky that such an improbable event occurred early in the Earth's history. It happened quickly, and it happened only once in the last four billion years.

But if this first incarnation of life was ushered in by God, then the early and singular occurrence of life on Earth makes complete sense. Since the Creator controls quantum effects to direct the world as He chooses, the creation of life may still have been natural. That natural process was controlled by God to make the impossible become possible. Life could have begun through one of the natural mechanisms hypothesized by scientists, with that mechanism being designed by God. This would allow man to one day discover how God brought life to be. Alternatively, this could have been a miraculous event where the first cell is made in an instant, initiating a biological Big Bang. We are a long way off from discovering an answer, though many creative ideas abound.

God made matter with the ability to form lengthy complex molecules that are ideally suited for life. We see this in the fine-tuning of the universe and by our very own existence. He also may have gone a step further and made a way for self-assembling molecules to arise naturally, built into the fundamental properties of matter. The universe could be fine-tuned for abiogenesis. Even with a natural process built in, it may be so unlikely to occur that it needs a divine quantum nudge to get started at just the right time. We have not discovered the process from which life began, but the fact that it has eluded us so long indicates it would be a delicate and rare occurrence hinging on the finely-tuned properties of nature.

Complex life-forms: Eukaryotes

The earliest known life on Earth were prokaryotes, which are microorganisms such as bacteria, that have a simple cell structure. Prokaryotes lack organelles (miniature organs within the cell) that more advanced cells have, like our own. Prokaryotes are single-celled creatures whose DNA, protein construction apparatus, and power generation all occur within the same mix of chemicals in a single container, the cell. These simple organisms were all that existed on the Earth until the emergence of eukaryotes, about two billion years ago.[20]

Eukaryotes are organisms such as protists, fungi, plants, and animals, whose cells have a complex compartmentalized structure. They can be multicellular, but single-celled eukaryotes exist as well. Multicellular eukaryotes didn't show up until just over a billion years ago. The defining feature of the eukaryote cell is that they all have organelles within them.[21] This includes the nucleus (the repository for DNA in long chains called chromosomes), mitochondria (the power plants of the cell), vacuoles (energy storage), and others specific to the cell type, such as chloroplasts in plants.[22] In all, there are ten or more different types of organelles in a typical eukaryote cell.

Like all life, all eukaryotes have descended from a common ancestor. This is a testament to the rarity of the evolution of complexity on Earth. Only once in 4 billion years of evolution did life learn to create energy producing organelles from creatures that lacked them.[23] This is why some have reasoned that complex life may be quite rare in the universe, as proposed in the book *Rare Earth*.[24]

Mitochondria may have been the first organelle to be incorporated in the first eukaryote cell. Unlike the origin of life, the origin of mitochondria is believed to be known. The basic principle is that at some point, one early cell was invaded by another as a parasite or engulfed in an attempt by one to ingest the other. In this unprecedented event, both survived. As these creatures co-evolved, their descendants eventually became symbionts, each requiring the other for survival. The smaller then became part of the larger as it lost all functions except that function the host cell needed it for.[25] The function it required was energy production.

The remarkable thing about eukaryotes is their capacity to store a much greater supply of DNA compared to prokaryotes. This is due to the limitation of prokaryotes to support merely a few thousand genes with the energy available to them.[26] A problem for them arises when the cell or its genome gets too large. To compensate for a lack of energy, some bacteria make multiple copies of their entire genome in order to create more energy processing proteins. But there is a limit to this, as the extra DNA is very costly in terms of energy.[27]

In the case of eukaryotes, the first symbiont became the energy producer, the mitochondrion. Mitochondria contain their own DNA and exist in all eukaryote cells. They are not parasites but essential components. Mitochondria solve the energy problem by containing only the necessary genes for energy processing, making their DNA quite small. This allows thousands of mitochondria to operate in a single eukaryote cell, providing an ample supply of energy. This extra energy is available to support a much larger cell and up to 200,000 times more DNA.[28] Without large genomes to work with, an organism gets stuck in an evolutionary rut. Bacteria are relatively unchanged over billions of years. Without the advent of the mitochondria, complex life would be impossible. According to experts on cell energy, "even aliens will need mitochondria".[29]

While eukaryotes engulf other cells quite often, this occurs rarely with other cell types. Bacteria are smaller and have a more rigid cell wall. When a foreign cell is engulfed, or a cell is invaded, the final result is that one or the other dies. This is why only once in the history of life on Earth, was a mitochondrion produced. All eukaryotes are descended from that first pair of symbionts with the descendants of the smaller becoming mitochondria. Prokaryotes have successfully advanced into eukaryotes only once in Earth's history.[30] We know this because mitochondria have their own DNA, recording the common descent of all eukaryotes. This represents another extremely fortunate and improbable event in the progression of life.

God's Role in the Origin of Life

Scientists are usually opposed to using acts of God as an explanation for any natural phenomena. And this is not necessary for scientific explanations of "how it happened". But when we ask "why it happened", we leave the realm of science and into philosophy and religion where invoking God to answer "why" is not only valid but, in this case, is quite necessary. Without God to help out the string of insurmountable odds that we find in science, our existence is difficult to understand. But with God, it all makes sense.

Some will cite the vast number of worlds in the universe as an explanation for life's ability to overcome the odds. Even though Earth-like planets may be rare, the sheer size of the universe allows for trillions of worlds, with a small percentage of those having the right mix of materials, a stable sun, and the right temperature for life. The anthropic principle is invoked: only on those few planets where conditions are right, life has originated, complex cell structure has evolved, and intelligence has emerged, will there be anyone to ask, "How did my planet beat the odds?"

While this argument has some merit in regard to why life exists on this planet, it does nothing to answer the question, "Why does life exist at all?" For without the remarkable properties of water, carbon, oxygen, and countless substances necessary for life, there could be no life at all, anywhere. Again we see how the universe is fine-tuned for our existence – no DNA, no life; no RNA, no life. There are countless properties of matter that are fine-tuned for life's construction. Without these bio-friendly properties of carbon chemistry, life could never begin anywhere, regardless of how many worlds you have to try on.

Besides fine-tuning, there is nothing to say that abiogenesis is naturally possible. We have not been able to find a mechanism by which life could arise from non-living material. The simplest cell type, prokaryotes, are vastly complex, requiring millions of DNA and RNA sequences and thousands of other molecular structures to be in place for the cell to survive and reproduce. We reviewed just a few of the required processes and saw just how complex they are.

The inner workings of the cell are like a computerized factory. The logical code in our DNA and the many mechanisms required for cell function are indistinguishable from a designed system. A very plausible conclusion could be that life is simply too complex to ever form by random processes anywhere in the universe given any amount of time.

Some interesting research into the mapping of the genetic code has revealed a few anomalies. While all life uses the same letters (nucleotides), the same words (codons), and the same twenty amino acids, there are a few microbial species that have mutations in the way that there genetic code is translated. In these rare instances, one of the stop codons have been reprogrammed to code for an amino acid. These code variants demonstrate a point – that it is possible for the genetic code to have different interpretive forms.[31]

Most biologists who study the origin of life expect that it originated by some form of chemical evolution on pre-biotic substances, starting with a much simpler genetic code. They would reason that the code became more complex over time as it evolved to its present form. But if this were the case, we should see much more variance in the genetic code, because changes in one line of organisms would have occurred differently than they did in other groups. But as it is, the genetic instruction set is so universal that a gene from a firefly can be inserted into a tobacco plant, resulting in a plant that glows in the dark. Even bacteria can take human or animal genes to produce a protein, as is commonly done for medical purposes. It is remarkable that such diverse forms of life can interchange genes, and they are still interpreted accurately. Though we have seen that some small differences have occurred in a few life-forms, this looks identical to what you would expect for the case of our current genetic code being the starting point, in full form, then a few mutations occurring in rare cases to produce the slight variants.

This supports the possibility that God created the first cell (or cells) in its entirety, or that He created several different cells or organisms at some early time utilizing a common genetic design. As in computer programming, software developers often use common subroutines that can be reused by different applications. So we see in the genetic code a very strong resemblance to a designed system, and the pattern of its variation best supports that conclusion.

God may have created one or a few initial forms of life or created a marvelous process by which life could be generated naturally. Either way God would be in control. Either way it's truly miraculous. The rare events leading to complex life would also be under God's control, only happening when and where He desires. Yet the method may one day be detectable to sentient observers, revealing the "how" of God's most amazing acts of creation. What if an ancient crystal were found that formed the perfect micro-structure for catalyzing the construction of RNA nucleotides, forming the necessary set of genes for self-replication? This would act as a sort of "AllSpark" for biological life, as occurred in the *Transformers* movie series for robotic life. This type of structure could not be explained by chance. This might be found in a naturally occurring substance or one of obvious intelligent design. A given phenomenon can be both miracle and completely natural because God is the inventor of the physical laws and the controller of quantum events. It will be exciting to see future developments in this field and if new fine-tuned properties are found.

The origin of life still has a lot of mysteries yet to reveal. But we can see that any conceivable origin would require an act of God.

Chapter Thirteen

Extraterrestrial Life and Intelligence

The notion of extraterrestrial life has gone from purely the realm of science fiction to a rapidly-growing area of research. While no alien life-forms of any kind have ever been detected, the speculation is that if it can happen on Earth, perhaps it has happened somewhere else as well. While this subject is very speculative, it represents an interesting topic to explore. Some thought experiments can give us insight into this question, though any real answers may be many, many years away.

When looking for alien life, the best bet is to look for DNA-based life similar to our own. We know that this configuration works. Given the complexity, uniqueness of many chemicals involved, and the difficulty to initiate the first replicating organism, it is most likely the only one possible. It is certainly the only one proven to exist.

People have long wondered if life exists anywhere beyond Earth. We have sent spacecraft throughout the solar system searching for clues of past or present life. While no life-forms have been found, microbial life might well be very difficult to detect. Searching distant solar systems for primitive life seems practically impossible, telescopes lack the resolution, and spacecraft lack the range by many orders of magnitude. So the search for life beyond our solar system is directed towards intelligent life with the technology to transmit radio communications.

The Drake Equation

Frank Drake proposed a formula for estimating the number of extraterrestrial civilizations in our galaxy. This has become known as the Drake equation. $N = R^* F_p n_p F_l F_i F_c L.$ [1] Where R^* is the average rate of star formation in our galaxy (new stars/year). F_p is the fraction of stars in the galaxy that have planets. n_p is the average number of planets that can support life per star that has planets. F_l is the fraction of those potentially habitable planets that eventually do develop life. F_i is the fraction of planets with life that develop intelligent life. F_c is the fraction of civilizations that develop technology that transmit detectable signals into space. L is the average length of time that such a civilization continues to transmit detectable signals before it becomes extinct or chooses to stop transmitting for some other reason. The product of all of these factors is N the estimated number of detectable civilizations in our galaxy.

The star formation rate in our galaxy is known to be about ten stars/year. It is believed that most stars in the disk of the galaxy possess planets. But the rest of the factors become more and more speculative as you go down the list. This has led to a wide range of estimates based on different points of view for the values of these factors.

Early estimates for extraterrestrial life were quite optimistic. Carl Sagan concluded that there must be millions of civilizations within our own galaxy. But after many years of searching the skies, no alien transmissions have ever been found. This has led to the

Fermi Paradox. This is a statement of the apparent contradiction between the estimated number of civilizations and what has actually been observed. The very fact that no alien transmissions have been detected leads us to believe that more conservative values for the Drake Equation factors must be valid. This suggests very low probability of life, an even lower incidence of intelligent life, making civilizations like ours exceptionally rare.

Estimates based on Kepler mission results indicate that Earth-sized planets in the habitable zone of their host star may be quite common. However, there are many requirements for life besides just temperature and planet size. But it seems with the astronomical factors being quite favorable to finding other civilizations, it must be the biological ones that are throttling back on the occurrence of intelligence in the cosmos. If life begins only rarely on habitable planets and even more rarely progresses to the point of intelligence, then the current non-detection of alien communications is quite understandable. This lends support to the proposed condition of the last chapter – that life originates very rarely and must be ushered in by God. This seems to have been done only once or else very sparsely throughout the universe.

Requirements for Habitability

For a planet to have life, many requirements must be met. Lacking any one of these could preclude that planet from hosting even the simplest of organisms. These criteria have been deemed necessary by experts in the field of astrobiology.

(1) Where a solar system is located in the galaxy has a lot to do with potential habitability. Stars near the galactic center are bathed in massive doses of cosmic radiation. This would likely keep any planet existing there sterile. But the farther from the galaxy center you get, the rarer heavy elements become. The outer regions of the galaxy are too metal poor for large terrestrial planets to form. So mid-range between the galaxy center and its outer rim is best, form-

ing a galactic habitable zone.

Any solar system forming with low amounts of metals, will form smaller planets. Smaller planets will be less likely to retain a significant atmosphere. Our solar system is actually something of an anomaly. It contains 25% more heavy elements than the average for solar systems in our region.[2] It may be that this over abundance of heavy elements is a requirement, or perhaps it just helps improve the odds.

(2) The type of galaxy in which a solar system resides is also important. Small galaxies have a limited supply of heavy elements. The same is true of elliptical galaxies which contain mainly old stars and have very little new star formation. The best galaxy type to find life is in a spiral galaxy.

(3) The right temperature is critical for life. This means that a habitable planet must be at the right distance from its host star, in the star's habitable zone. This is generally defined as the band of area around the star that liquid water could exist somewhere on the surface of a planet. In our solar system, only one planet exists within this zone. This zone is defined by various people differently, but a more liberal estimate of this distance range is from 0.95 to 1.15 AU for a sun-like star.[3] This yields a narrow band where life could be found.

(4) When to find a habitable planet is as important as where. Rocky planets could not exist in the early universe. The Big Bang produced only hydrogen and helium, so other elements did not exist until they had been manufactured by the first generations of stars. These element factories would disperse their products through supernova explosions at the end of their lives. Only after about 2 billion years would enough heavy elements be available for life and planet building.[4] Heavy elements continue to be produced by stars, progressively increasing the material available to form planets. Our solar system didn't form until about 9 billion years after the Big

Bang.

(5) Stars come in many sizes, each having very different properties. The larger the star is, the shorter its lifespan. Life requires long periods of stability for evolution to occur. A star only 50% larger than the sun will have a lifespan of only 2 billion years. This is too short a time period for animal life to evolve, assuming the same pace of evolution as on Earth.[5]

Smaller stars have longer lifespans and are more numerous. But the habitable zone for these stars are too close to the star for free rotation. Any planets in the habitable zone would become tidally locked to the host star, so that one side always faces it. This results in a hot side and a cold side and only a small region on the border between night and day that is just the right temperature. But any atmosphere on this planet would freeze out on the cold side, as well as any water. So these planets would be airless and dry, not at all suitable for life.

(6) A large moon might be required for any planet to have animal life. A large moon stabilizes the rotation axis so that it does not drift too far out of alignment with its orbital axis. This keeps the seasons mild. Plate tectonics would not continue for long without tidal influences of a large moon. Though if the body in question was a moon, rather than a planet, the host planet may provide the tidal forces necessary. Plate tectonics is an important mechanism for trapping excess carbon and for continent building. More importantly, the tidal flexing induced by a large satellite keeps the planet core from solidifying. The strong magnetic field of a planet with a liquid metal core protects it from the charged particles raining down from space. Without this magnetic shield, radiation would flood the planet's surface.

Given all of these requirements, we see another reason why life appears to be rare. Many separate criteria need to simultaneously be met. This is the hypothesis presented in the book, *Rare Earth*, by

Peter Ward and Donald Brownlee. Though this may be disappointing to those hoping to make contact with extraterrestrials, it may be quite fortunate for us. Any race we might encounter would almost certainly be millions of years more advanced than ourselves, given that we are just barely space capable. We need only look at our own history to see how less advanced civilizations fare when encountered by more technologically advanced peoples.

Galactic Colonization

The idea of colonizing our galaxy may seem far fetched now, with spaceflight technology still being very limited. However, if our civilization lasts another thousand years or so, this might not only be possible, it might be inevitable.

To be certain, the distances between stars are vast. The nearest star to Earth, Alpha Centari, is 4.3 light-years away. The fastest spacecraft ever flown, as of this writing, is the New Horizons space-craft en route to Pluto. It reached its peak velocity after a gravity assist from Jupiter, traveling at 23 km/s (14 mi/s).[6] This speed was attained with conventional rockets and gravity assist maneuvers. In comparison to the speed of light, this is on the order of 0.01% of light speed, or about one ten thousandth of the speed of light. At this pace, it would take at least 43,000 years to make the journey to even the nearest star.

Propulsion technology is continually improving, and there are many yet-to-be-developed technologies that are feasible with known physics. It is not hard to imagine spacecraft ten times faster in the near future. And if any advanced form of nuclear propulsion is developed, spacecraft at least a hundred times faster will be possible. Some of these concepts include nuclear electric propulsion, nuclear pulse drives, and thermal nuclear rockets. Though even at 1% of light speed, it would take 430 years to get to Alpha Centari.

Even faster travel rates may be possible, but perhaps there could be physical limits preventing higher speeds. Things like interstellar gas and dust could pose a problem, for instance. But even if an advanced civilization is technologically limited to traveling at only 1% of light speed, it could still colonize a galaxy. Such a race could develop a way of freezing or otherwise suspending the crew during flight. Or medical breakthroughs could advance the lifespan of the species to the point that such long flight times are survivable. Another feasible way to achieve interstellar travel is a generational ship. This would be a city-sized spacecraft that would contain everything a civilization would need. The original travelers would not survive the trip, but their descendants would eventually arrive at the destination. Robot probes could also be used and almost certainly would precede any colonization attempts.

Even at these reasonably modest speeds, an advanced race could spread throughout the galaxy in about 10 million years. This result simply follows from the fact that the galaxy is 100,000 light-years across, and at 1% of light speed, this is the time it would take to traverse it. The time to establish colonies and send new travelers might be almost negligible compared to the travel time. Even a slow rate of establishment, like the colonizing of North America, only took a few hundred years. So for a slow rate of establishment (say 400 years), it takes about 20 million years to colonize the galaxy. This is still a brief period for a galaxy that is over 13 billion years old.

Given the short period between the rise of an intelligent species and the colonization of a galaxy, it could be expected that if an extraterrestrial race is detected, then they likely inhabit the whole of their home galaxy. Modern humans have existed for only tens of thousands of years, an extremely brief period in comparison to the length of time to colonize the galaxy. So the single planet phase of a civilization's development is very brief. One might ask, "Why would the alien race have such galactic ambitions?" Any alien race would likely comprise differing cultures within its own species. It would only take one with the will and means. The large travel times and communications lag would isolate individual colonies. Once established, they could develop distinct cultures of their own.

Given the tendency to fill a galaxy, rare incidence of intelligent life might well be favorable. Otherwise species would frequently clash, and younger civilizations would be at risk of being exploited or exterminated by more developed civilizations. This might be regarded as a desired property of the created universe. In His wisdom, the Creator could have ensured that intelligence occurs only rarely to prevent dangerous interspecies interactions. With intergalactic distances so great (millions of light-years), crossing from one galaxy to another seems very unlikely or at least very rare.

The universe is fine-tuned for life and is huge, so it seems likely that the intent is for more than just one occurrence. But rare occurrence would still be favorable. Even if intelligence exists in only a thousandth or a millionth of all galaxies, that still adds up to millions or billions of civilizations across the visible universe. Perhaps SETI, could one day have success not in our own galaxy, but by finding alien communications from distant galaxies. The combined signals of an entire galaxy of communications or some other technological activity may be detectable from great distances.

Chapter Fourteen

Questions, Problems, and Misrepresentations

With science having had so much success in explaining the nature of the universe, you might think that we have a pretty firm grasp on things. Think again. Follow any popular science literature for a few months and you find numerous articles that conflict with each other, or expose errors of prior work. Observe technical science journals, where new research and theories are submitted for presentation to the scientific community, and the conflicts and disagreements abound all the more. One gets a sense of just how uncertain many truths we would like to accept really are.

If we are to attempt to answer the deep question of origin, we need to also understand the uncertainties involved in the data that we are using to make our judgments. Much has been said about what is known, but attention must also be paid to what is unknown. The more we discover about the universe, the more questions we have.

All of science is constrained by the uncertainties in the data from which it is based. This is also true of the position taken in this book. To be objective, we must consider the issues that are still unresolved for the *Designed to Evolve* principle.

Unsolved Mysteries of the Universe

Our understanding of the universe has blossomed over the last century. We have gone from thinking that the Milky Way was the extent of the universe to the realization that it is actually one galaxy out of billions. We once thought that the universe was ageless but now know it to be approximately 13.8 billion years old. Yet many questions remain. So many that by listing them our ignorance is truly exposed.

What is Dark Energy?

According to the basic form of general relativity, it would be expected that the expansion of the universe would be decelerating due to the pull of gravity. However in 1998, came the discovery that the universe's expansion was actually picking up speed. Some unknown energy field was imparting a force in opposition to gravity. Current observation indicates that the best representation of this force is in the form of the cosmological constant, a type of force that is constant per unit of volume. Since it is proportional to volume, its effect was small when the universe was younger. But as the universe expands, the influence of this force increases.

The energy density of the universe is about 73% dark energy. Though its contribution can be estimated, we have no idea what is actually causing it. Nor do we know if its strength (per volume) will change in the future. There are a number of possibilities, but no conclusions can yet be drawn. This is one of the most vexing problems in physics. It is likely we will be searching for clues to this for years to come.

What is Dark Matter made of?

This problem is older than the problem of dark energy and is nearer to a solution. However at the present, it is still unanswered. When the speed of stars are measured, it would be expected that the stars nearer to the galactic center would orbit faster than stars in the far reaches of the galaxy. This follows directly from Newton's law of gravitation. But the actual measurements show that the stars orbit at approximately the same speed for quite a large range of distances. Their speeds are also larger than what would be expected given the visible matter present in the galaxy. The best explanation for this is that there must be some unseen mass dispersed throughout the galaxy to provide the needed gravitational tug to keep all the stars in their present orbits.

This unseen dark matter must only weakly interact with normal matter or else it would have condensed inside the galaxy long ago. Also, it must be the largest component of a galaxy, outweighing stars and planets ten to one. All galaxies and galaxy clusters are dominated by the effects of dark matter.

It is hoped that particle accelerators will soon produce the particle or particles that make up dark matter. If they can produce it in the lab, then its properties can be discovered. This would be a great triumph of physics; its discoverers will most certainly be awarded the Nobel Prize.

What is the nature of gravity at the quantum level?

Since the discovery of quantum mechanics, theorists have sought a way to link quantum mechanics to general relativity. Without a theory of quantum gravity we are left guessing as to what exists inside of black holes and at the earliest moments after the Big Bang.

Why is there an imbalance of matter over antimatter in the universe?

Einstein showed us that energy and matter are equivalent and one can be converted to the other. When energy is converted to matter particles, an equal amount of antimatter is also produced. Shortly after the Big Bang, the universe consisted of pure energy and no matter. So it would be expected that when the expanding universe cooled enough for particles to form that equal amounts of matter and antimatter would be produced. Fortunately this was not the case, or else the antimatter is hidden somewhere, since we cannot detect it. For the universe is largely devoid of antimatter with only trace amounts found.

Where did this imbalance come from? Is there some law of physics that in some way favors matter? Or is the matter dominance only local and some very distant regions are actually filled with anti-matter and missing normal matter? Efforts to create and trap neutral antimatter atoms may one day lead to detailed analysis of the properties of antimatter. If any deviation in its properties as compared to normal matter are found, this could point to new physics governing antimatter and lead to an explanation for its scarcity in the universe.

How big is the universe?

The age of the universe is known, but the size and shape of the universe is still a mystery. Some sources will quote the size of the observable universe, but this is just the part of the universe that we can see. The entire universe is much larger. Most evidence suggests that the spacial geometry of the universe is very nearly flat. As stated earlier, a flat universe does not require it to be infinite. There are several 4-dimensional manifolds that are topologically flat but finite in volume. However, a flat universe would open up the possibility that the universe is actually infinite in size. The very fact that we do not even know if the universe is infinite or finite shows our overall ignorance on this subject. If the universe is truly infinite, this will be difficult for us to discover, since an extremely large size may be impossible for us to distinguish from infinite size.

What caused the Big Bang? What caused inflation?

There is no consensus theory to explain what initiated the Big Bang. It could be said that God did it, by the establishment of finely-tuned initial conditions. Alternately, a natural mechanism may be responsible, since God most often uses the forces of nature to achieve His goals. He may have established laws of nature with this mechanism in place, causing the universe to begin. But in either case this represents a truly miraculous event.

The same question can be posed for inflation. Early in the universe, a period of inflation is thought to have occurred, when the dimensions of the universe were increased many orders of magnitude. The cause may be related to dark energy or some other unknown phenomena.

Both of these questions are unknown and may be without a natural solution that we are capable of learning anytime soon.

By what mechanism did life originate on Earth? Are alternate biologies possible?

This is an active area of research right now. It seems there are many proposed explanations but no consensus at the time. But whatever the mechanism, it represents a truly rare event requiring many ideal circumstances to simultaneously be met. The fact that all life on Earth descended from the same common ancestor suggests that life's incarnation only happened once in the 4.5 billion years of Earth's history. But the fact that it happened so early is something of a puzzle.

Many have cited life's early appearance as evidence that life begins easily when the right conditions are met. But if that were the case, it would be expected that it would have happened multiple times here on Earth. And if the various domains of life each had unique ancestry that would be the obvious conclusion. But, in fact, they do not. All Earth life shares a common design and common descent. So either there is simply no trace of the other unique ancestral lines left now, or it actually is an extremely rare and difficult occurrence.

Some have cited panspermia to explain life's origin. Panspermia proposes that life on Earth came here from space, not in spaceships but as microbes hitching rides on asteroids. Proponents of this idea claim this as an explanation for how life began so soon on Earth against the incredible odds. But for life to originate in another solar system and come here would have taken millions of years travel time. It is unlikely any microbe could survive that long in space. Panspermia would not provide any additional time if life came from another body in our own solar system, because all the planets in our solar system formed at the same time.

When considering the possibility of finding life on other planets, it has been asked if their chemistry would be expected to use DNA like ours does. Or are there other chemistries possible? We have never found any other life, nor been able to produce it synthetically in the laboratory. So testing this idea will be difficult. But given the complex nature of life and the uniquely complex properties of carbon chemistry, it seems unlikely that another significantly different molecular formulation could support the complexity required for life. Any alternate formulation of life might be so different that we might not be able to classify it as living. And it may not possess the same levels of complexity necessary to ever allow for intelligence. This is certainly an unanswered question, and any suggestion of alternate biologies is still highly speculative.

Much still to learn.

So as can be seen from this sampling of unknowns, we have a lot to learn. There are many basic features of our universe that are still a complete mystery. When people make claims that science has figured it all out, or that there is no room for God in nature, they are speaking their opinion, not fact.

Issues within the Designed to Evolve Model

As I have embarked on a personal exploration of the details of our universe and evaluated every piece of evidence and every argument, I have formed a world view that I believe to be the best approximation of reality. The purpose of this book is to present the data to you so that you can evaluate the merits of this position your-self. The fundamental evidence for the existence and for the role of God as Creator is well established. So with this premise as an axiom, I believe that future discovery will only enhance this posi-tion. But, as with any model that aims to extrapolate to the limits of the known data, there are some issues left outstanding. I want to be sure that these are not ignored, so that a complete assessment can be made. These issues may impact some details of this model but not the central tenets.

Is the "Fine Structure Constant" constant?

The fine structure constant represents a ratio of several funda-mental constants including the speed of light, the electric charge, and Planck's constant. Many studies on this subject have placed increasing restriction on any potential variation in the fine structure constant. A recent study concluded that the fine structure constant is indeed constant to at least 0.001% throughout the visible universe.[1] This supports the position that the fundamental constants of nature are indeed constant.

However, a new study has been released that claims to have found statistical evidence for possible variation of the fine structure constant at the 1/10000 of a percent level between opposite sides of the visible universe.[2] Though this is a tiny difference, if confirmed, it would throw the laws of physics upside down. Constants are expected to be constant. However, textbooks need not be rewritten yet, since this claim will need to be investigated with more sensitive instruments. The variation that was reported was below the accuracy of the individual measurements taken. As is often the case with bold claims, time usually exposes the flaws in such sensational results. Most other research supports no variation in the constants of physics.

But if variation of the fundamental constants were found to be a property of the universe, and the universe was found to be infinite in size, this could detract from the fine-tuning argument. An infinite universe and variation in the values of constants would provide for the possibility of a selection effect as in a multiverse. But for this position to be valid, both unlimited constant variation and an infinite universe would be required. The size of the universe is still unknown, and the true extent of the universe may not be known for many years yet to come.

Is the universe infinite in size?

This was listed as one of the greatest mysteries of science. We do not know if the universe is finite or infinite in size. This leaves the door open for those who wish to speculate about things that can not be seen, such as a multiverse. But such speculations do little to detract from the significance of fine-tuning. The universe could well be infinite in size yet finite in age and have identical physics throughout. Such a state would still be fine-tuned and still require a creator.

But if the universe were somehow proved to be infinite in size, infinite in age (contrary to Big Bang theory), and the constants were proved to vary though space, then this would provide a possible natural explanation for fine-tuning. Of course, there is no proof for an infinite universe, varying constants, or anything preceding the Big Bang.

On the other hand, if the universe is found to be finite in size, the chances of a multiverse become vastly slim. The only natural explanation for the fine-tuned state of the universe is by means of a selection effect within a multiverse. Any determination that the universe is finite in size would eliminate this alternative explanation for fine-tuning, leaving creation by God as the sole theory of origin.

Where are the early plant fossils?

Small soft plants do not fossilize easily. To date, none have been found predating the Cambrian. This seems to run counter to DNA evidence for the divergence of vascular land plants 700 million years ago. Some could rightly argue that DNA dating is better at predicting order than actual dates. They might then say that the lack of plant fossils at this early time is simply because there were none. This would obviously run contrary to order of creation given in Genesis if the modern definition of plant is used. It was this apparent out-of-order condition of plants that first prevented me from recognizing the Genesis creation order in the fossil and geological records. The DNA evidence is less frequently cited since it is more uncertain than rock-solid fossil evidence. But DNA can give a clue to future discovery of fossils.

This problem is not a real problem, just a minor detail. While I believe that the plants cited in Genesis do include primitive land plants, it does not need to. The modern definition of plants is quite restrictive, but in ancient times it was quite broad. The biblical usage of the word plant could have included plants, fungi, lichens, and algae. So even if early plant fossils never turn up, these other forms of immobile life are firmly documented in the fossil record prior to the Cambrian explosion.

Another alternative is that perhaps I have the dates wrong. An alternate way to correlate the creation account might have day five starting in the Devonian period, which is known as the "age of fish." Before this time only very small animals existed and in relatively few forms. The Devonian saw a vast diversity of new types of fish and other aquatic animals of great size. This would have allowed day three to include the ages up to the Silurian or even early Devonian when many plants and even trees are known to have existed on land. This would require a very short day four, but this is still a possible solution. Perhaps future fossil discoveries will make this more clear.

Evolution is a hard sell to many theists.

There is no doubt that evolution poses some difficult theological questions. Many details of creation have a long history and deep traditions of being interpreted a different way. This is simply because we didn't have any understanding of evolutionary thought before Darwin's time. We can now see that the phrase, "Let the earth bring forth...", used in Genesis could be a depiction of God commanding the forces of nature to produce these things over time, evolution being one of God's creative tools. The order of appearance in the fossil record is in excellent agreement with the Genesis creation account. But even if we are able to view the creation account in light of evolution, there are other issues.

The real difficulty in accepting evolution into theology is not the creation account; it is multitude of new questions it raises. Many things that were assumed from traditional interpretations now come into question. Now we ask, "Were Adam and Eve literal people? Did they also evolve? Does this imply death before the fall of man?" These are legitimate questions that have to be addressed when evolution is applied to theology. These are not easy questions, and there are no easy answers. Just as many other difficult theological questions surround the topics of predestination, the Nephilim, and interpretation of the many events described in Revelation.

While these questions may be outside the scope of this book, here are some opinions on them, not as an absolute theological stance but purely speculation as to how these things could be understood. Adam's name in Hebrew means "mankind", but it could be that the word for mankind came from his name. I believe that the nature of the Genesis stories and genealogies support a literal man named Adam. Though he may not have been the first humanoid, he was the first human to call on the name of the Lord. It may be that he was the first human given a soul by God. Without a soul, any pre-existing physical being would not be fully human. Adam may have been given the required traits to make him physically human and then received a soul to make him spiritually human as well. The creation of Eve from Adam's DNA would have been a miraculous event ensuring any special traits he had were part of her as well. While Adam and Eve were preserved by God so that they would not die, at least until they took the forbidden fruit, their ancestors and the rest of the animal kingdom did suffer death. There are certainly other ways to interpret the scriptures relative to known science and evolution, but this is my opinion based on my belief in the authority of scripture and how God works in the world.

As stated earlier, progressive creationism can yield much the same results as evolution. Though the evidence currently favors evolution, perhaps future discovery could lend support for the special creation of various forms of life at specific times in history. My view of evolution does not exclude that possibility. Since God is in control at all times, He could make more drastic changes any time He desires. If this were the case for major animal groups, then evolution would look more like progressive creationism and could be more limited in scope than the existing evidence suggests. Ultimately, the principles of this book still apply whether you choose to believe in theistic evolution, progressive creationism, or anywhere in between.

Other Objections.

Some may allege that the conclusions of this book are fringe science, or an attempt to smuggle God in the back door. Others will object because they believe the Bible requires that the Earth and universe are young.

The notion that God is control of the natural forces and designed them to make the world orderly and discoverable to men is not a new idea. This viewpoint was held by the founding fathers of science. Through this understanding they had faith that the laws of physics could be discerned though scientific investigation. Theistic evolution has been around since evolution was first proposed. Even Darwin admitted that God could have created the first primitive forms of life and set evolution in motion.[3] There have always been more theists that believe in evolution than atheists.

Additionally, the science reported on and used in this book is well established. Speculative theories and new unverified hypothesis are not used in shaping the designed to evolve principle. With patience we must wait for full verification of any scientific finding before it is incorporated into our world view.

To those who believe that the Bible requires a young Earth, I would like to point out that there are a lot of assumptions made to arrive at that point of view. A different position on a few of these assumptions and you arrive at old earth creationism. The two are not that different theologically. When two options are possible from one set of data (the Bible) and one of these is excluded from another set of data (modern science), then it logically follows the position not excluded is correct (old earth creationism). This is especially evident, since this method independently predicts many findings of science.

Not all will accept the findings of this book. But disagreement cannot be made by classifying it as fringe science; the science used is well established. I encourage anyone with any doubt to verify the facts from neutral sources and read other books on this subject. Of course, there are a wide spectrum of viewpoints. Many good books on this subject can be found in the general references at the end of this book. In the end, I hope that this book will help you find a deeper understanding of God and the natural order.

Speculative Science and Fringe Theories

Just as tabloids often spread half-truths and blatant lies about Hollywood life, fringe science can confuse the public about issues of science. Speculative theories by prominent scientists can also spark confusion due to misrepresentation of the uncertainties involved. There are many speculative theories floating around that attempt to answer various questions in physics. Some of these may one day prove to be valid, but looking at history, most will likely be replaced as new data is obtained. That is the nature of science. Theories that are well proven and have stood the test of time are unlikely to be supplanted. But newer findings, that have not yet had independent verification or theories that have not had time for predictions to be observed will be subject to scrutiny with the possibility of being discredited. Some more speculative ideas are based on untested, unverified, or unsupported theories, like the idea of a multiverse. These ideas creep into scientific literature and often make wild claims. While they may represent some fascinating thought experiments, they most likely have no basis in reality.

When forming a view of reality, the level of certainty of each piece of information must be accounted for. Every opposing opinion has some arguments made for it, or else it would not have supporters. But the accuracy and certainty of these ideas vary from one opinion to another. So it is prudent to take a critical eye to any new finding. Here are some of the speculative ideas out there that pertain to the subject matter of this book.

String Theory / M-Theory.

This theory is often the starting point of most multiverse theories, grand unified theories, and theories of quantum gravity. But it is as yet unproven. The basic idea is that all fundamental particles are made up of microscopic strings that vibrate within various extra dimensions of space and time. String theory covers a broad range of theories. Some variants propose as many as eleven dimensions. M-theory regards the one dimensional strings as slices of two dimensional membranes that vibrate in eleven dimensional space-time.

The problem with these theories and their many variants are that they can be tuned to fit almost any situation. So rather than showing the true nature of microscopic entities, it may only be a mathematical construct tuned to match any available data. The extreme flexibility of the theory makes it difficult to generate testable predictions. And since there are so many variants, many of them are in conflict with each other. There are currently many physicists that do not support string theory. It still lacks the proof required for broad acceptance.

Alternate Cosmologies: The Cyclic Universe.

As we discussed previously, the cyclic model of the universe has effectively been demoted to the least likely universal model. But a recent study claimed that echoes of previous Big Bang/Big Crunch cycles could be detected in the cosmic microwave background.[4] The authors of this study hoped to reinstate the cyclic universe model and, in so doing, claim the universe to be eternal.

An eternal universe could be a detraction to the creation model proposed in this book. With the universe always existing, there would be no scientific basis for it to need to be created. This essentially elevates the universe itself to the status of an eternal creative force, using infinite time and randomness to generate the favorable conditions for life. This would not disprove God's existence, but it would remove some of the basic evidence that now provides strong support for a creator.

So does the rational person, upon hearing of this study, immediately alter their world view, the model that they use to understand the whole of reality? Not at all. One cannot simply change your own world view at every whim, or you will have no viewpoint at all. Science necessarily explores all possibilities and, by this process, will generate many incorrect results in the short term. But in the long term, these errant findings will be culled out through the peer review process and further study.

This is just what has happened with this cyclic universe model. New analysis has found the basic premise of this cyclic universe study was flawed, and the features they claimed to be evidence of past Big Bang cycles were simply features of the singular Big Bang that is the basis of the standard model of cosmology.[5] The standard Big Bang is maintained, and the universe still has a finite age.

It is important to consider the probability for error of any findings and not just the error bars supplied by the authors. Sometimes the conclusions are based solely on speculative assumptions. Just because the study is published in a peer reviewed journal does not mean it is scientific certainty. These journals are the sounding boards of the scientific community, where ideas are shared for review by a broader audience. The scientific community itself does not accept any theory until it has had independent validation. Even with this critical process, errant ideas will occasionally gain broad acceptance for a time. Science makes no claim of perfection; it is a work in progress, always testing and reevaluating, getting an ever better picture of reality.

Alternate Biologies.

If you read astrobiology news articles, there are occasionally references to alternate biologies. But there is little to back these speculative ideas up. Our DNA-based life is uniquely complex and ideally suited for the task of biology. Some astrobiologists hope for the existence of an alternate chemistry of life because this would increase the odds of finding extraterrestrial life-forms. Others might support this position because they don't like the idea that our chemistry is special or fine-tuned for us.

In a typical example of hype from the science media market, comes the claim of arsenic-based life. The claim, however, was not for a novel form of life but one sharing common descent with the rest of Earth life that happens to utilize arsenic in its DNA as a substitution for phosphorus. This is no small claim. It implies that perhaps an alternate form of life could emerge somewhere else that does not use phosphorus at all but instead uses arsenic in its DNA.

It took little time for experts to cry foul on this claimed discovery. They criticized the methods and the conclusions. Many cited the inherent instability of arsenate as precluding it from replacing phosphate in DNA. Experts in the field also presented alternate, less spectacular explanations for the results of the experiment.

Eternal Inflation.

There are many ideas describing what exists beyond the limits of detectability. One could conservatively presume that what lies beyond is more of the same. But more speculative ideas also exist. Since there is presently no way of determining what lies beyond, we do not know.

One of these speculative theories proposes that the universe is infinite in size. It postulates that the inflationary epoch during the first few moments of the traditional Big Bang theory stopped only locally, forming the visible universe as just a local region of the larger multiverse. While inflation stopped here, it continues unabated throughout most of the cosmos. Occasionally, it stops in other places also throughout infinite space. This is known as eternal inflation. This is one proposed mechanism for creating the alternate universes of a multiverse.

Though this is a popular model among some cosmologists, it is very speculative. There is no known mechanism to drive inflation eternally. It represents a direct violation of the Second Law of Thermodynamics. Also, this theory does not explain how inflation got started in the first place. Eternal inflation was covered in detail in chapter five, describing the various multiverse theories. Anyone can dream up an untestable idea. This is a prime example of untested theories propagating through the media. Many books have proposed that some sort of multiverse explains our just-right universe. But these claims are mere speculation. It makes little sense to base one's view of reality on such conjecture.

Caution needed in regard to uncertain concepts.

As shown throughout this book and in many others, the body of well-supported theories and data show that the simplest, most straight forward answer is that the universe was created by God. The vast body of well-established science attests to that.

There are large volumes of theoretical science that is as yet unproven. Many of these theories are in conflict with each other or with more established models. In the end, some of these may be turn out to be true, but most will not.

Beware of popular news postings which propagate every fringe idea. Science media outlets will embrace every spectacular headline, hyping up any intriguing story. But these are often based on a single source with unverified results. Science moves at too slow a pace for most enthusiasts. So popular media jumps on every new idea, with little regard for the validity of the claims.

The Internet is a haven for fringe science. Science and physics blogs are full of wild and incorrect information. While entertaining at times, they often have little basis in reality. Yet some people seem to formulate opinions based on what they read on these blogs. It is always important to consider the reliability of the source. If information is not backed up with legitimate sources, it is likely errant.

There are many books out there claiming the exact opposite conclusion of this one. But I have not seen any with a credible or complete argument. Most are based on speculative ideas that have their foundation in unproven or untested theories. These are often authored by famous scientists with strong anti-religious sentiments, using their notoriety to hustle their opinions. Others like to shoot down the straw man, using the fact that the young earth creation model can be easily disproved. They ignore the fact that other more scientifically robust creation models exist. Some just resort to name calling and insults toward people of faith. This seems to be an attempt to use peer pressure to bully their opinion onto others. These books like to equate faith to delusion or dementia. They claim that intelligence requires their point of view.

Some of the bitterness towards religion may stem from past experience of being treated poorly by religious individuals with anti-science or anti-evolution stances. This is unfortunate. Such poor behaviors certainly are not endorsed biblically or academically. Those who resort to using insults are exposing the lack of strength for their argument. This is true regardless of the subject matter or the position being disputed.

There are many strong viewpoints on this subject. This leads to many strongly slanted publications. Many scientific theories are presented solely to uphold the modern assumption of atheism. When you hear a new piece of information in the news or in a science journal that conflicts with your individual viewpoint, consider the source and potential for error. Give it some time; many new, world-changing theories are often proved false by future examination.

Chapter Fifteen
Principal World Views

There are a number of ways to view the whole of reality. A complete view of the cosmos requires both the scientific and the spiritual perspective. Even if your viewpoint denies one of these, that denial defines your viewpoint. A world view defines not only the way you view the world but also your values, goals, and lifestyle. It represents the filter by which you interpret the world around you. Here is a brief overview of the major world views as applied to origins.

Naturalism, Agnosticism, Deism

Naturalism or atheism is the belief that no god of any kind exists. It is often claimed that science can explain our existence through theories of the Big Bang and evolution without invoking any supernatural influences. Many naturalists are unaware of the scientific evidence for God or prefer to explain away this evidence by chance. Some will support ideas like that of a multiverse to back up their view. Though the evidence is lacking for these theories, they expect that the evidence will one day be found. This could be called a "science of the gaps" position since the currently available and proven science points to a creator.

Naturalism in the United States is supported by about 16% of the population.[1] This figure likely includes those who would consider themselves agnostic. An agnostic presumes there to be no god, because they believe that God's existence cannot be known. Those who do not care whether or not there is a god would also be considered agnostic.

Deism holds that there is a god who created the universe but that this god had no further influence. The deist's god has no involvement in the affairs of man and bears no resemblance to the God of the Bible. Einstein was probably the most notable deist. He believed the laws of physics to suggest a creator, but he thought this creator to be too great an intellect to have any interest in us.

Young Earth Creationism

While the previous positions deny the involvement of God in the universe, the opposite is true for Young Earth Creationism. To hold this position requires rejection of many areas of science, namely the time scales and processes of cosmology, geology, evolution, paleontology, etc. This is largely due to the belief in a young age of the Earth and universe.

This viewpoint holds to traditional beliefs for the interpretation of the biblical account of creation. It also maintains that the days of creation in Genesis were literal twenty-four hour days and that the age of the Earth and the universe is a little over 6,000 years. The age is based on the work of James Usher, who in 1650 dated the first day of creation as Sunday, October 23, 4004 BC. In the *Annals of the Old Testament*, he tabulated dates for creation and other events from biblical genealogies. He used the latest scientific and historical information available in his day to fill in the gaps.[2] Though science has changed, this viewpoint on creation has not. It has become doctrine in some of the most conservative religious institutions.

Old Earth Creationism without Evolution

As its name implies, it accepts the age of the universe and the Earth as determined through science. Old earth creationism covers a wide range of viewpoints. The most notable distinction between these views are whether or not they include macro evolution as part of God's creation toolbox. For many, the stigma of evolution invokes the idea of a godless process that is incompatible with common ideas about creation. It seems too haphazard a way for an omnipotent Creator to form beings in his image. For many, the traditional understanding of the Genesis creation account seems incompatible with evolution, even if a day-age interpretation is held. This leads to an old earth view that includes the special creation of life. Though macro evolution is rejected with this view, some degree of micro-evolution is usually accepted to explain common changes to pathogens, the success of breeding, and other small changes in species over time.

Some old earth creationists still support literal twenty-four hour days in Genesis, and allow for an old Earth by the implied gap between the creation of "the heavens and the earth" in Gen 1:1 and the beginning of the first day in Gen 1:3. This has been known as "gap theory".

Intelligent Design

Old earth creationism provides the philosophical basis for intelligent design. Intelligent design is not a religious or philosophical view of reality in itself but a scientific method. It is the scientific investigation into phenomena that may be best described by having an intelligent cause. This can include a variety of theories such as the origin of certain properties of the universe, or properties of the Earth, but most often is used in the study of biological systems. This can cover a wide range of viewpoints, since the identity of the intelligent architect is not stated. Even evolution is not excluded from being an accurate picture of biology, but it is subject to scientific determination given that supernatural causes are also accepted as possible. So in this context, intelligent design could include theistic evolution. However, in practice most intelligent design proponents do not support macro-evolution.

With researchers accepting the possibility of supernatural causes in nature, intelligent design provides a framework that could allow for the discovery of greater truths than is possible in a scientific arena that adheres to the assumption of atheism. Yet to work, this would require strict adherence to critical review and openness toward a wide spectrum of philosophical viewpoints. This is especially true regarding the inclusion of natural explanations presented as a check to any intelligent design claim. As it is, there is far too little invested into science that evaluates alternatives to purely naturalistic explanations when it comes to theories of origins.

Progressive Creationism

A variant of old earth creationism is progressive creationism. By this, life on Earth is created in full form, by special creative events throughout geological time. This may have been done either with just a few events mirroring the six days of creation or a continual string of creative works. This position is used to explain the progressive nature of the fossil record, while maintaining special creation of life-forms. The appearance of common descent can be explained as God using the DNA from preceding organisms as a starting point for creating new forms of life. An example of this is seen in the creation of Eve; God forms her from one of Adam's ribs.[3]

Most old earth creationists accept micro evolution to one degree or another. This allows for adaptation within a species and for changes in viral and bacterial pathogens that are well documented. Also, adaptation of breeds of animals is permitted. Some supporters of progressive creationism will also accept speciation within a higher level group such as the family, order, or phylum. The Cambrian explosion and other explosive bursts of new life-forms in the fossil record are often cited as evidence for progressive creation. Detecting differences with such scenarios and full macro evolution would be difficult given that both theories share many of the same features.

Evolutionary Creationism and Theistic Evolution

These two titles represent two names for nearly the same model. In addition, there is another name for this; biologos, suggested by Francis Collins, director of the Human Genome Project.[4] The many different names represent the desire to shed preconceptions of evolution and creationism that are not intended by this position. For simplicity, theistic evolution will be used here. Theistic evolution is the belief that God used evolution as a tool to create the variety of life on Earth. This perspective accepts scientific findings about the common descent of life.

The frequent issue with theistic evolution is separating the atheistic stigma that often comes with evolution. But this view on evolution implies that the Creator uses evolution as a tool for creating species. What is really at the heart of why some believers are rejecting evolution is the simultaneous promotion of atheism. This is a legitimate concern, especially in schools. Evolution needs to be taught, but there needs to be effort made to not promote atheism in the process. This could be accomplished by including a study of the philosophical views on evolution, as well as teaching the limits of evolution, such as the origin of life and any problems and dilemmas within the theory.

Though all of the creation viewpoints agree on the role of God as creator, they differ widely in their approach of melding science with theology. For theistic evolution, the ancient accounts of creation in scripture are often taken as allegory, where spiritual truths are described through a hypothetical story. These descriptions are not regarded to represent actual history but a teaching about spiritual principles. While this viewpoint provides an easy explanation for any discrepancy with science, it is not a theologically acceptable manner of interpretation for many believers.

Theistic evolution maintains that God created the universe to generate life. He built into the properties of matter the ability to support life and the ability for life to be generated through some natural process. Theistic evolution sees God as getting everything started at the Big Bang, then stepping aside and letting the physical laws do the rest. He would have set everything in place, knowing each individual that would eventually be born, though He would interact with the world only rarely through miracles. His more common interaction would be spiritual, through personal revelation.

There are many who believe that theistic evolution represents the best picture of how God created life on Earth. Though many versions of this belief exist. Some supporters of theistic evolution would embrace far more interaction by God than the mainstream version allows. This could be seen as a separate creation model altogether and is summarized in the next section.

The Designed to Evolve Principle

This line of reasoning finds that biblical texts need not be allegory to accommodate acceptance of evolutionary science. This view recognizes scripture as an accurate account of history through proper context and when the use of metaphor is considered for the time scales involved. This represents a more theologically sound methodology. As was concluded earlier in the analysis of Genesis chapter one, the biblical creation account is a remarkable prediction of the future scientific discovery of our time. This text describes progressive stages of creation that are very much in line with modern geology and the fossil record.

The designed to evolve principle could include progressive creationism, with evolution working to modify the basic body plans originally created. This would be the case if some limiting factor was ever found that restricts the extent that evolution can have on life, or that the Cambrian explosion was too abrupt to ever have originated naturally by evolution alone. As of yet, a limiting factor has never been proven to exist and may be difficult to prove, even if it is there.

Theistic evolution fits the designed to evolve principle with the major difference being that the biblical creation account is taken to be a historical account, not simply allegory. The order of events described in Genesis one are intentional and accurate. This viewpoint sees God as being in continual control of the process, even when natural mechanisms are able to describe it completely. In essence, natural mechanisms have spiritual purpose. And all natural phenomena can be guided through God's continual control at the quantum level.

The origin of life is another subject apart from evolution. But it is of great interest to any world view, and may provide an answer to the extent that evolution has been used by God. While many details of evolution are well understood, the origin of life is still quite unsolved. But there is scientific reason to expect that the Creator was required to provide a mechanism or a divine act. This was discussed previously in chapter twelve on the origin of life. Within the designed to evolve principle, the origin of life mechanism could have been built into the design of matter and the laws of nature. This represents another level of fine-tuning, since matter would be specially designed for the initiation of life, as well as for its continuation. Or it could have been a special creative act on the scale of the Big Bang as used in creating the material universe. In this case, God would have created one or several initial life-forms, from whence all life descended. The uniformity and intricacy of the genetic code lends evidence for the designed nature of life's internal programming. In either event, God is ultimately responsible and in control at all times.

This is the core of the designed to evolve principle: that God is in complete control of the universe, from beginning to end. Natural forces are his tools; the natural laws are his design. The Bible represents an accurate account of history and is inspired by God. Whether progressive creationism, theistic evolution, or somewhere in between, these models fit into the designed to evolve principle. This postulates that all events are controlled by God at the quantum level; if life evolved, it did so with God guiding it every step of the way.

Choosing a World View

When judging between some of the various world views, it seems prudent to make use of available scientific information. If it were shown scientifically that biological features and DNA sequences of life are better understood by instantaneous creation than by common descent, then that viewpoint would be accepted. As it is presently, there is limited evidence for this position in science. Evolutionary theory is quite successful at predicting many kinds of patterns found in the DNA of various lifeforms. But if there was a theoretical alternate to evolution that was a truly superior explanation and predictor of patterns in DNA, it would have to be accepted in mainstream science. This is not to say that such profound conclusions would not face ferocious opposition, but if the theory were true, it would be proved out in time. Since the Bible does not clearly specify whether or not life evolved, either in the whole or in part, it seems that this question aught to be left to scientific investigation. The basic premise that God is discoverable in his creation, as stated in Psalm 19, provides confidence for this method.

Though scientific investigation has largely supported evolution, there has been little investment into alternate theories. But a promising path of study for an intelligent design researcher may be in the origin of life. As there is no obvious natural mechanism, this leaves open the possibility of a divine "Big Bang" in biology. That is, if God created the first ancestor or ancestors of all life in full form, then evidence of this might be discernible somewhere in the genetic code of life. On the other hand, if a natural pathway to life's origin is found, there will likely be incredible fine-tuning examples within this pathway that will reveal God's handiwork. There remain many opportunities to discover more examples of how the universe and the properties of matter are fine-tuned for life.

Given the long line of evidence presented in this book and many others, the atheist position is shown to be lacking the capacity to explain existence at any level. Some naturalists rely on their gut feelings or belief that their viewpoint will one day be proven. But this represents a sort of blind faith that they often criticize of religion. Logic is not on the atheist's side; reason runs contrary to this position.

As for me, the physical evidence for God's existence, biblical predictions, revelation in scripture, and my own personal experience leave no room for doubt. This evidence is sufficient to mandate the existence of God and his role in the creation of the universe. A world view lacking God would be logically incomplete. Creationism holds that God created everything. It is the when's and how's that separate all the remaining world views into their respective flavors of creationism. In consideration of the evidence, it is up to you to choose the view that fits best.

Chapter Sixteen
Summary of Evidence

As we survey the evidence, a complete picture of the universe is revealed. Behind the natural forces, there is the guiding hand of God. The universe evolved from a hot, dense initial state to host the vast sea of worlds in the present age, all according to God's design. The cosmos and the mechanisms behind its evolution reveal convincing evidence for God's role in creating and sustaining the universe.

Even biological evolution is incomplete without God. Whether God progressively created life at many separate stages or gradually guided evolution from microbe to manatee, it is all His design. Many atheists have used evolution as an excuse to deny God. But this is a fallacy of logic. There is no reason that evolution could not be another device in God's tool kit. An evaluation of the biological processes that make evolution possible reveal a complex system of micro machinery; this must be in place before life or even evolution can function. There are two sides to every coin, one side cannot exist without the other. Understanding reality is incomplete if either the physical or the spiritual side is neglected.

The forces of nature are the tools of God. The universe is

designed to be discoverable so that we can learn the details of creation and use this knowledge to better our lives. The *designed to evolve* principle is not a new concept, but its basic features were presumed by the founding fathers of science. Newton, Maxwell, Galileo, and many other great minds believed that the remarkable degree of order in the natural world was God's design. They recognized that God worked through nature to accomplish his will.

Here is an overview of the evidence for God as creator of the universe. At no other time in history has there been so much scientific support for creation as there is today. This will continue to increase as science advances. There are many areas of research that are on the brink of exciting discoveries that will add to our knowledge. The more we discover, the more God is revealed.

The Temporary Universe

The First and Second Laws of Thermodynamics prevent energy from being created or destroyed and require that energy always takes the path to a less useful state. As energy is converted from one form to another, it does great things: it powers stars, it powers cell growth, it powers our bodies and our machines. But once used, there is no return to the previous state without a supply of more unused energy. This shows us the time-limited nature of the universe. It began with an immense store of useful energy and has been steadily burning through it ever since. Someday that energy will be exhausted; the universe as we know it will end. The universe cannot continue as it is forever due to these two laws of thermodynamics.

Big Bang theory follows from general relativity and has been validated by extensive observational evidence. This theory states that the universe is not eternal but had a beginning. The Big Bang is a great description of creation – all the energy in the universe originating in an instant, and all the matter in the universe forming moments later. Time and space also begin at the Big Bang, showing how God is truly outside of the temporal and physical dimensions, as these are merely aspects of the created universe. This is just the sort of creative event you would expect from an omnipotent God. A universe with a beginning needs something to create it, a first cause.

Throughout modern science, there have been theories aiming to explain the universe's existence without invoking God. These theories use the modern assumption of atheism as their basis. This often means that some exception to the First and Second Laws of Thermodynamics must be made in order to allow nature to be capable of creating itself. But time and time again, these theories are embraced for their agreement with the modern assumption and then disproved later when more data was available. The steady-state universe and the cyclic-universe models violated the fundamental thermodynamic laws only to be disproved after years of promotion by cosmologists. Now the multiverse is the latest prodigy of the modern assumption, also in violation of these laws.

The First and Second Laws of Thermodynamics have held up to centuries of scientific investigation. They are called laws because they are so well proven and result from simple mathematics. It makes little sense to disregard these foundational precepts in favor of speculative theories, such as the multiverse. The straight-forward application of these laws and the evidence in support of Big Bang theory suggest the action of a benevolent Creator.

The Fine-Tuned Universe

As we look at the equations that describe the expansion of space since the Big Bang, we see that remarkable precision is required in setting the values of the universe's fundamental parameters, such as the gravitational constant, the density of the universe, and the initial force of the expansion. If any of these properties differed from their measured values only minutely, life would not be possible, anywhere in the universe. Also, the same conclusion is seen for many of the other various constants in physics and chemistry. Seemingly unrelated branches of science are connected by relationships between these constants that are extremely fine-tuned for life. Some have said that this position is unimaginative. If the constants were different and carbon-based life were not possible, perhaps another form of life would be. But these parameters are so sensitive to modification that many, if different, would prevent the existence of any complexity at all. It is hard to imagine any form of life in a universe devoid of stars, or lacking elements heavier than helium, or without any atoms at all. This is an undisputed fact: the constants of physics and the initial parameters of the Big Bang could not have been different to any significant margin without excluding any possibility for our existence.

There is no natural explanation for the values of these parameters. There is no scientific reason for this fortunate state of physics. This would not be expected of a random assignment of these values, but suggests that they were selected intentionally by the Creator, to allow for life. Speculating about higher level structures beyond our universe (such as a multiverse) only moves the fine-tuning problem to a larger scale.

Quantum Mechanics

Due to the uncertainty principle, the universe is not deterministic. It is not rigidly following natural laws that completely determine what events will ensue. Random effects, that cannot be predicted by theory, occur at the quantum level. If God occasionally takes control of these microscopic processes, it would be undetectable to us. Through this mechanism, God can maintain complete control of all creation, at every point in time, yet never violate a single law of physics. Quantum mechanics not only allows for control by God, but it requires that He maintain continuous involvement to guarantee any final outcome. Otherwise random quantum effects would drown out any course of events set at an earlier time.

By this same mechanism, He has provided humans with free will. By allowing the soul to interact with the brain at the quantum level, the soul wields control of the mind. Yet this control is immeasurable, being masked by the uncertainty principle. The human mind is a remarkable device. No understanding of it is complete without recognition of its spiritual component. Without the soul's influence, we would consist of only computer processing and chemical reactions. Through modern theories of consciousness such as Orch-OR, the brain's dependence on quantum computation is revealed. This theory shows how consciousness is lost when anesthesia causes quantum entanglement to be disrupted in the microtubules within the neurons of the brain. Without this quantum effect, the soul loses contact with the brain and consciousness is lost. Quantum fluctuations within our neurons allow for decisions to be made that depart from the logical processing of data, providing a gateway to the soul. Without the soul and this quantum process, we would be mere robots, and we would lack free will.

A created realm would be expected to be quantized. Think of a computer system. There is a minimum level of precision on the screen, the pixel, and a smallest unit of information, the bit. The fact that the world is made up of indivisible particles with only certain energy levels permitted, shows resemblance to a created system.

Throughout the ages, people have ascribed the forces of nature as being "acts of God". The Bible contains numerous accounts of miraculous events being carried out through natural phenomena. Storms are calmed, seas are parted, plagues of wind, hail, insects, etc. Literally anything can be accomplished with complete control of the quantum realm. The physical laws guide the weather, the cosmos, and life, as a sort of auto pilot. Yet the control of natural forces by God, as recognized by our ancestors, now can be quantified and explained. A built-in control mechanism at the quantum level represents an ideal way for the Creator to maintain dominion over the universe, while still providing orderly laws of nature that are discoverable and exploitable by created beings.

The Equivalence of Information and Energy

The equivalence of energy and information reveals insight into the nature of God. This equivalence principle represents a possible creation mechanism. Any source of infinite information would necessarily be a source of infinite energy. Since God is defined as having infinite knowledge, He is therefore all powerful. The equivalence of energy to knowledge provides a way for the universe's initial store of low entropy energy to be created. This method of creation is one that only an all-knowing Creator with control of the quantum realm could employ. "The Lord by wisdom founded the earth; by understanding established the heavens" (Proverbs 3:19 ESV).

Predictions in Genesis One of Future Discovery

A number of predictions about Earth's early history can be identified in the biblical creation account of Genesis chapter one. These predictions of future discovery about our planet's ancient history were made thousands of years before the existence of modern science. Only in our time, are they able to be tested. Remarkably, the predictions and the order of occurrence are in excellent agreement with modern scientific findings. Here are some examples:

1. The universe is not eternal; it has a finite age. (See Gen 1:1)
2. The surface of the early Earth was dark. (See Gen 1:2)
3. Early in Earth's history, it was covered in a worldwide ocean. (See Gen 1:2)
4. The Earth lacked continents in the earliest epochs. (See Gen 1:9)
5. The formation of a supercontinent. (See Gen 1:9-10)
6. The recent arrival of man. (See Gen 1:26-28)
7. The formation of the Earth was not instantaneous but was progressive, each step taking time. (See Gen 1:1-31)
8. The correct order of key events in the history of the Earth are predicted:

i. The creation of the universe, including the Sun, Earth, Moon, and stars. (See Gen 1:1)

ii. The early state of the Earth's surface. (Worldwide oceans, darkness.) (See Gen 1:2)

iii. The initial thinning of the atmosphere. (Light on the Earth) (See Gen 1:3-5)

iv. The reduction of water in the atmosphere. (Separate the waters.) (See Gen 1:6-8)

v. The formation of continents. (Dry land appears.) (See Gen 1:9-10)

vi. The appearance of plant life on land or the development of photosynthesis. (See Gen 1:11-13)

vii. The evolution of a transparent atmosphere. (Lights visible in the sky) (See Gen 1:14-16)

viii. The appearance of sea creatures. (The Cambrian Explosion) (See Gen 1:20-21)

ix. The appearance of birds. (Extant since the time of the dinosaurs.) (See Gen 1:21-22)

x. The appearance of modern land animals. (Placental mammals after dinosaurs.) (See Gen 1:24-25)

xi. The appearance of modern humans. (See Gen 1:26-28)

The creation account of the Bible has provided countless generations with information about Earth's history. Now, recent discoveries have confirmed the details of this account. This provides a strong endorsement of the Bible with very specific predictions that cannot be dismissed as mere coincidence. Through this evidence the Creator's identity as the God of the Bible is verified.

Evolution

When we get to the subject of evolution, there is much contention on this issue. Atheists will often use evidence for evolution as an argument against God. Many believers find the subject of evolution incompatible with their ideas about how God created life on Earth. But as we have shown, the Bible is not that specific on this issue. It is not hard to see how evolution could be implied in some passages. Though there are some valid arguments that challenge evolution, the majority of scientific findings presently support it. This is not a problem for the design argument, since evolution could be seen as a tool in God's creation toolbox.

While there is strong evidence for evolution, there are also anomalies in the theory that need to be addressed. The fossil record does not read quite so gradual as evolutionary theory implies. The rapid appearance of new animal phylum in the Cambrian Explosion is just one example. Other issues appear in the study of DNA. The cell processes its DNA like a computer running a program. The complexity and logical nature of the genetic code resembles an intelligently crafted system. This same coding system is shared by all life on Earth. Computer engineers at Apple may not have been first to invent copy and paste. Within the cell, there is a built-in ability to copy and paste data, delete sequences, and substitute single bases in the genomes of offspring. This ability for change is inherent to all life, but it is not self-evident. Many features of evolution simply do not make sense unless life is *designed to evolve*.

Origin of Life

Evolution requires that life already exists to work. Evolution does not work on molecules or non-living matter because evolution requires self-replication. While many aspects of evolution are well supported by DNA and fossil evidence, the origin of life is less understood. Though ideas abound, there is no scientific consensus as to how life began. Some have argued that the minimum level of complexity required to allow for self-preservation and self-replication is so great that the purely natural origin may be impossible. The fact that all life has descended from the same common ancestor is testament to the rarity of life's origination. It only happened once on

Earth in 4.5 billion years. Yet it happened early in Earth's history at the optimal time for the eventual development of advanced life. For the rare incarnation of life to occur only by chance, at just the right time, seems too good to be true. This does not necessarily preclude the existence of a natural path to life, but it suggests that it would need to be either built into the nature of matter by God, or that God specifically created the first life-forms by some unknown mechanism. We may one day discover how He did it through scientific investigation, but this will by no means discount this miraculous act.

The long string of fortunate events that led to intelligence on Earth need an explanation. Most biologists claim that evolution proceeds without any direction or goal. But the progressive advance of life over the eons shows otherwise. The development of mitochondria, which supplies energy to all eukaryote cells (plant & animal cells), only happened once, as all eukaryotes have a common ancestor. Chloroplasts also have a singular origin, providing all plants with photosynthesis. Many exceedingly rare events have punctuated biology, allowing complex life to evolve on Earth. It seems unlikely that these rare events are just coincidence; the guiding hand of God is evident throughout life's history.

The genetic coding system of life is exceedingly complex yet is common to all life. If simpler forms of this coding system existed, you might see how it could have evolved from a simpler form. But that is not the case; all life uses this same complex coding system to program its DNA. The only variants differ by a singular change to the amino acid assignment of a single codon. These are obvious degenerate forms that mutated from the common system. Our genetic code is so consistent that genes from a firefly can be substituted into a tobacco plant to make it glow in the dark. This common coding system, that has no evolutionary origin, provides a strong signal for design. There exists no better explanation for the genetic code than direct creation by God.

Morality and Purpose Need a Creator

In addition to all of the physical evidence for God, there are many philosophical reasons as well, frankly, too many to cover here, since that would be outside the scope of this book. But a couple will be discussed due to the significance of this subject.

Moral perception has little evolutionary advantage. But it is this perception that shows us the flaw in the atheist's stance. For if you presume atheism, that there is no god, all that exists is a natural fluke, an accident, a temporary condition. Then some truly horrific conclusions about life can follow. To follow this thinking to its conclusion, life is meaningless and empty, no more than chemical reactions and interactions of particles. Self worth is a fantasy. Morality of any kind is meaningless. The only admirable goal would be to optimize the propagation of your own DNA. If it was necessary to steal, to lie, or to kill to advance your own position, that would be prudent, according to a strictly naturalistic view. That is, if there were no built-in moral mechanism. But the fact that even the most convinced of atheists would not likely hold these views, is due to our God-given moral instincts.

These moral instincts do not make sense within an atheistic world view. They would be an evolved trait that is to the detriment of the individual. To accept the idea that there is no god, one must accept these horrifying conclusions and discount moral instincts as an evolutionary blunder. Only with God does morality have any validity. Only with God does life have purpose. If there is no God, there is no soul. This would render life as being no more than chemistry.

The Belief of Billions is not Delusion

There have been prominent atheists who have stated that belief in God is akin to delusion, as if to say that those believers must have one foot in the door of the mental ward. If that were true, the world would need a mental ward with a multibillion patient capacity to contain this fanatic scourge. The truth is that the vast majority of the faithful are productive, competent members of society. They include doctors, engineers, housewives, scientists, teachers, nurses, bus drivers, loggers, miners, and people of every occupation. Many of these people are highly intelligent; some are experts in their field. These people are quite capable of logical reasoning and making sound decisions.

Yet if there were no God, these people would all be truly insane. For nearly all of these would acknowledge, not only personal interaction with God, but a personal relationship with him as well. This would represent the ultimate imaginary friend. But since these people are indeed rational, productive members of society, they are clearly not insane. The Friend they have is not imaginary; He is real.

For the believer, the greatest evidence is not any presented so far in this book. It is the voice of God Himself that has echoed in each believer's soul. Why could these rational, intelligent people never be convinced of the atheist's view? It is because of their personal experience with God that leaves all other arguments mute.

Chapter Seventeen
Overview of the Theistic Universe

A theory of everything is a sought after goal of particle physics. Three of the fundamental forces – electromagnetism, the weak nuclear force, and the strong nuclear force – have been unified under a single comprehensive theory. But one of the fundamental forces is presently left out, gravity. To include all four fundamental forces into a single theory would provide the core to a theory of everything. Another aspect of this complete theory of everything would be the unification of general relativity and quantum mechanics. Currently, there is no theory to describe situations that include both relativistic and quantum effects simultaneously.

While these pursuits are certainly admirable, the goal of this book is to present a theory of everything of a different sort. This is a complete picture of all reality, including the spiritual and the physical. It is not an exact scientific specification but an overall picture the physical universe and how it relates to the spiritual side of reality. This requires a cohesive philosophy of how the spiritual realm is interwoven with the physical realm.

The scientific evidence for a creator is overwhelming. The Bible has been verified by the correlation of its predictions with the scientific understanding of Earth's ancient history. By accepting the validity of the Bible, the existence of God, and the reliability of science, a cohesive world view can be formed. This view is no substitute for the Bible, nor is it a scientific discourse, but an understanding of reality shaped by these sources.

The Beginning of Time

It is of no surprise that God knows the future. Countless biblical examples of this are available. The fact that time is a local property according to general relativity, intricately linked to the universe itself, indicates that the one dimension of time and the three dimensions of space were also created as part of the universe. This gives us a picture of God, forming the universe not at a moment, then allowing time to move forward, but as a 4D structure where time is no different (to Him) than the other dimensions of space. The fact that one of those dimensions reflects time for us is simply the artistic style employed. This is why He sees the future as if it has already happened. God is not limited to our temporal frame, but He can act at any location in time or space without regard to the normal flow of time. The laws of nature are chosen to create a world with beautiful continuity that is discoverable to created beings.

Think of an artist crafting a mural on a long canvas that depicts chronological scenes in succession. He could have painted the last scenes first or in any order, yet the observer views it chronologically with proper respect for cause and effect from scene to scene. The reason for the flow of time, gravity, or any other consistent pattern is due to the creative style of the artist. To any character within the mural, the artistic style will seem as immutable physical laws. God chose to make these laws consistent so that they are discoverable to us. He may also leave some areas unfinished so that we can participate in shaping elements of the story.

Since time and space are constructs of this universe, the Creator is outside of these devices. There was no "before" the Big Bang, because time began at that instant. Time is an invention of God, as is space, that expanded rapidly after the Big Bang and continues to expand as you move forward through time. Space was not created empty but full of energy in a state of very low entropy. Acting at the quantum level and having infinite knowledge, only God could have created this very energetic, very orderly, low entropy state at the beginning of time.

On the universal canvas, the Big Bang marks the beginning. The laws of nature and initial parameters of space-time were the commands of the Supreme Architect. Time begins, the universe expands from near nothingness to thousands of light-years across in the blink of an eye. The expansion was propelled by the laws of physics set at the beginning, yet guided at the quantum level by God. Due to the uncertainty principle, no specific final outcome is predictable. For any future intended goal, it would be necessary for the Creator to at least periodically, but perhaps continually, intervene in the course of events. The galaxies we see today result from quantum fluctuations in the local energy density during those first moments following the Big Bang. To choose the shape and properties of individual galaxies would require control of these quantum events occurring long before galaxy formation began. God created the universe with a plan and with a purpose; fulfilling this purpose would require his continued involvement with creation.

Planet Earth and Life

Over the eons, the Creator tends his universe, as atoms form, then stars and galaxies, then heavy elements, then planets, as the universe develops according to his laws. We can imagine the Creator surveying his creation, perhaps searching it, for just the right place to begin a marvelous work. This is the picture we get from Genesis: "The earth was without form and void, and darkness was on the face of the deep. And the Spirit of God was hovering over the face of the waters." (Gen 1:2 ESV). Finding a water world, whose formation had occurred by his design, He chose it and wielded influence over its evolution.

In the first two days of the creation account, there is no mention of life, though it may be implied by analogy, as light is often a metaphor for life. It was certainly early in Earth's history when life first appeared. Once on Earth, life was a transforming force. Here we see another tool of God. These early life-forms were not just evolutionary stepping stones to future life, but they were chemical processing machines preparing the Earth's environment.

Was a self-assembly mechanism for life built into the nature of matter? Or was life constructed as a new creation within creation, starting a biological Big Bang? The commonality of the genetic code of life indicates that this code did not morph gradually from a more primitive state but was a basic feature from life's first incarnation. Its complexity and versatility indicates a high level of intelligence, with all the necessary components available from the beginning to allow for advanced forms of life like us.

Certain primitive life-forms were required to condition the Earth. These life-forms transformed the soil, added oxygen to the atmosphere, and provided food for other creatures. Along with the other forces of nature, life and evolution provided a divine tool kit for fulfilling God's purpose. The progressive formation of new life-forms may have been very gradual or more stepwise; both situations are found in the fossil record. In either case, God is in control, ushering in new species at their proper time. Built into the machinery of life, into its DNA, is the ability to adapt and change to best suit its environment. This is a remarkable property that allows for the great diversity of life on Earth. Life is a magnificently designed invention that can reproduce itself and evolve through changing its internal programming. It is astounding that all the diverse forms of life with their own unique programs, all use the same genetic machinery.

The many steps on the evolutionary road to complex life, seem at odds with possibility. But God makes the impossible possible, as each roadblock is overcome at precisely the right time. God is long suffering and patient, as time is no boundary for Him. The ages of Earth's history are only a week by His measure. Most of the many varieties of life have existed for millions of years. But modern humans have existed for merely thousands of years. There were pre-human forms, but these lacked the capacity for creative thought and the ability to know God. Then the first male and female of the modern human lineage were endowed with a soul, completing the first man and woman. The Bible describes the creation of Eve as a task in genetic engineering, as she is constructed from Adam's own DNA (from his rib). Perhaps he was also engineered from the DNA of what we now interpret as his genetic ancestors. From the DNA of creatures that were merely animals, new cognitive beings were made, having within them the breath of God. They could know God and discern right from wrong. They could discover; they could create; they could learn how to love.

Life on Other Planets

With all the innumerable worlds within the created realm, there seems no reason why God would not have performed this task many times on many planets throughout the universe. God is not limited by time or space, as He is everywhere at all times. But this prospect is certainly speculative.

What about primitive life occurring in places that could never support complex life-forms? Some theorists have proposed that primitive life should be much more common than complex life and that complex life is far from an inevitable outcome.[1] Should primitive life be found to have originated on some inhospitable world, this would be evidence for a built-in mechanism of matter to generate life, especially if this second genesis also uses DNA. This would be another example of fine-tuning on the grandest of scales.

We have never found any signs of life on a world other than our own. Due the great distances, there may be no way to detect primitive life even if does exist elsewhere. It seems that intelligent life may be more detectable, since high power radio transmissions could spread across vast distances through space. We have been scanning the sky for decades in search of alien signals, but no signals have been found. However, the distances between intelligent civilizations could be so vast that there has not been time, since these civilizations began broadcasting, for speed of light transmissions to reach us. In the absence of detectable alien transmissions, we are likely the only inhabitants of our galaxy, and the nearest major galaxy is several million light-years away.

Human History

Over many years of human history, God has revealed Himself to humans through His messengers, the prophets, the apostles, and His Son. These revelations, along with historical accounts were recorded in books, providing instruction and knowledge of God to all generations. These books would one day be collected and incorporated into the Holy Bible. The Bible records many actions of God: miracles which are often orchestrated through natural forces, quantum effects, and possibly the bending of the natural law. Through these accounts we know the nature of God. While all knowing, all powerful, and perfect in conduct, He is compassionate and forgiving of His created beings who have erred. Through the Old Testament law, the flawed human character was exposed. But as a gracious Father, He guides His children to do right, helping them to lead productive and content lives, teaching them how to get along and care for others.

When the Earth was ready to fully know God, He sent His Son, the human incarnate of Himself, to show the extent of His love for humanity. Willing to suffer death in order to establish a relationship with each person, He offers the gift of eternal life to those who accept it. In raising His Son to life again, God demonstrated His power over death and the ability to raise his followers to life again as well. He desires that all will choose to follow Him and have a personal friendship with Him.

The Present Age

As we have better learned to work together towards a common good, our ability to create and discover have advanced. Only through strong moral influences could humans be able to restrain their selfish, carnal nature. Benefiting from God's guiding hand, humanity has discovered the knowledge of science and how to use it for invention and technology. But in our arrogance, we have forgotten the One who gave us these abilities and taught us to use them. But God foreknowing our nature, provided a witness of Himself in the very fabric of His creation, so that in our advancing knowledge, we would not be able to forget the God who gave us life, nor miss out on a relationship with Him.

It is now in our time that we see the many proofs of God's role in creation. Only now have the predictions of future discovery been proved out, as revealed in the Genesis creation account. These details of Earth's ancient history were recorded thousands of years before science could discover them.

We have also discovered the delicate balance within the laws of physics. The initial parameters of the Big Bang and the fundamental constants of physics are extremely fine-tuned for life. We can even learn about God's character from studying His creation. Due to the consequences of the uncertainty principle, it is shown that God does not stand back and watch, but He must be continually involved in the affairs of the universe to guarantee any desired result.

Heaven and the Future

The early findings of science challenged many assumptions about the physical world. But as time has passed, the evidence for a created universe has increased steadily. Many theories that once claimed to supplant the need for a Creator have since been discredited. As time moves forward, this trend will certainly continue. This does not mean that there will not be those who will be determined to deny God. There will always be people of divergent views. But it can be expected that scientific evidence for the Creator, his characteristics, and his role in creation will continue to increase.

The Bible foretells of a time when the Earth, as we know it, will end. At this time those who are dead will rise and face judgment. Some will be granted eternal life. These are the ones who have accepted the gift of life, having a personal relationship with God through His Son, Jesus Christ. Those rejecting God, will be separated from Him eternally. This is not revenge but necessary to keep evil out of heaven. Who would give a stranger or an enemy the keys to their home? Entrance into heaven is no different; this is God's home.

The future offers the possibility of discovering life or even intelligent life elsewhere in the universe. Though we may never detect it, even if it does exist, due to the vast scale of space. It is quite possible that there are countless inhabited worlds. If God gave man a soul, could He not have given souls to other beings as well? Perhaps heaven is already populated with multitudes of those who lived during earlier epochs of the universe. Though we do not really know if any alien beings exist, perhaps this will one day be determined. Many questions are presently unanswered for us, so there will be much to discover in the life to come.

Journey's End

Ultimately, when examining the great wealth of scientific discovery of our time, each one of us must choose between two world views. One claims all that exists is an accident, the result of an inevitable outcome of chance on infinite scales of time and space. The other claims that all is created for a purpose, by an all powerful, benevolent Creator who desires for us to discover Him.

The data points overwhelmingly to purpose. That purpose cannot be explained away by chance but requires an intelligent Architect. Only when we recognize this fact does any discipline of science make sense. Even evolution fails as a theory without a designer to set the wheels of life's magnificent machinery in motion.

The fine-tuning of the laws of physics is a testament to the universe's created nature. Multiverse theories cannot detract from this condition. Lacking evidence, some theories are contrived to justify the assumption that there is no god. But the evidence for fine-tuning is so well documented, that it has spurred a bit of desperation on the part of some authors of atheistic literature. This shows the merit of the fine-tuning argument.

When the biblical account of creation is read in context, no contradiction with science is encountered. In fact, a stunning degree of accuracy is found when testing its predictions against the science of our age. The details of Earth's ancient history were revealed to man thousands of years before science could verify these claims. Now confirmed, this is one of the most remarkable verifications of biblical authenticity.

This journey began to discover the origin of our world, and through it we also discovered the nature of God. Not only is the Creator's existence verified by science, but His identity as well. Through scientific examination, His acts and His character are revealed. Just as it is written: "The heavens declare the glory of God, and the sky above proclaims his handiwork. Day to day pours out speech, and night to night reveals knowledge." (Ps 19:1-2, ESV). Certainly, scientific revelations of God are no substitute for the Bible, His personal revelation to humanity. But they attest to the legitimacy of the Bible and its relevancy in the twenty-first century.

It is most amazing to consider that the God who made the universe, in its vast complexity and extent, desires a personal relationship with each one of us. The One who occupies the whole universe still sees us and cares for every individual. The journey to discovery is exciting, and the end of this journey is to find truth. But it is really only the beginning, to know about God. To be complete, we need a relationship with Him, through His Son, Jesus Christ. Only then, can we one day join Him in heaven and discover wonders that are beyond all comprehension.

Glossary

abiogenesis: The formation of life from non-living matter. It usually refers to the origination of the first life-form or life-forms on Earth or on some other planet.

archosaur: An ancient clade of reptile that predates dinosaurs and crocodilians.

antimatter: Every normal matter particle has a corresponding anti-matter particle that is equal in mass but opposite in electric charge and a few other parameters.

atheist: Someone who professes to believe that there is no god of any kind.

black hole: An object of sufficient density that gravity overcomes all other natural forces so that it colapses into a singularity. The boundary of the black hole is known as the event horizon, beyond which nothing can escape (not even light).

billion: $1,000,000,000; 10^9$

Boltzmann factor: This factor gives the relative proportion between two states based on the energy level of each state. The Boltzman factor results when the system is in thermal equilibrium.

CMB (cosmic microwave background): This thermal radiation was emitted about 380,000 years after the Big Bang when the primordial gases cooled enough for neutral atoms to form. It continues to travel through space today emanating from all directions.

codon: A sequence of three nucleotides in the genetic code. There are sixty-four codons that indicate whether to start, stop, or add a specific amino acid during construction of a protein.

cosmological constant: The constant in Einstein's theory of relativity that represents the strength of dark energy is known as the cosmological constant.

creation science: A term for non-standard science prepared by non-scientists that aims to back up the belief that the Earth and all species of life were created about 6000 years ago.

dark matter: Unseen matter made of particles that only rarely interact with normal matter, except by gravitation. Its existence is implied by the motions of stars within galaxies and the motions of galaxies within clusters, suggesting that this material far exceeds the amount of normal visible matter in the universe.

dark energy: A gravitationally repulsive force believed to be responsible for the acceleration of the universe's expansion.

deist: Someone who believes that God created the universe but denies that God has ever had any interaction with humans.

deuterium: A stable heavy isotope of hydrogen that contains one proton and one neutron in its nucleus. (Normal hydrogen only has a proton in its nucleus.)

DNA (deoxyribose nucleic acid): The double stranded molecule responsible for the storage of genetic information in all known life.

energy: The capacity to drive change in a system. It can take many forms such as, heat, light, motion, mass, gravitational potential, pressure, or the strength of a field.

entropy: The disorder of a system, or a measure of how many different ways a system can be arranged microscopically, and still have the same macroscopic properties or thermodynamic state. The lower the entropy, the more orderly the system and the greater the potential to do work.

enzyme: A protein that acts to chemically alter other biochemicals. It may be used to break molecules apart or to join them together. Each enzyme has very specific target molecules it acts upon.

eukaryote: All organisms that store their DNA in a nucleus are eukaryotes. This includes all plants, animals, fungi, and protists.

fauna: The various types of animal life existing in a particular region or geological period.

flora: The various types of plant life existing in a particular region or geological period.

field: An energy gradient that fills the space between particles. Some common force-inducing fields are gravitational fields, electric fields, magnetic fields, nuclear fields, or theoretical fields such as the inflaton field or a dark energy field.

fine-tuned: Something that has been intelligently manipulated for a desired purpose, such as the strings of a piano tuned to hit the correct notes. This term is also used for the properties of the universe and the constants of physics that are fine-tuned to allow for life in the universe.

fusion: A nuclear reaction where two atomic nuclei join to form a heavier atomic nuclei. Only certain combinations of atoms under the right conditions can join in this way.

galaxy: A large collection of stars, dust, and gas that are gravitationally bound, orbiting a mutual center.

half-life: The length of time that it takes for half of an unstable substance to decay. All radioactive substances and unstable particles decay logarithmically, such that half of the substance is gone after one half-life, then half of what remains is gone after another half-life, and so on.

heaven: The term heaven can refer to the universe, the sky as viewed from Earth, or the spiritual realm where God dwells.

heaven and earth: The entire universe.

hydrocarbon: Any molecule that contains both hydrogen and carbon, as in all biochemicals.

inflation: Cosmic inflation is a theory postulating that the universe went through a brief phase of rapid expansion shortly after the Big Bang.

inflaton: The theoretical particle responsible for cosmic inflation.

isotope: An isotope is a variant of a particular chemical element that is differentiated by the number of neutrons in its nucleus. There are two non-radioactive isotopes of carbon, carbon-12 and carbon-13, having six and seven neutrons respectively. Carbon-14, the most stable of the radioactive isotopes of carbon, has eight neutrons and a half-life of 5700 years.

Kelvin: A unit of measure for temperature, equal in size to a degree Celsius, but starting at absolute zero (the lowest possible temperature, 0 K, -273.15 °C, or -459.67 °F. This is the total absence of thermal energy.)

light-year: The distance that light travels in one year. About 9.5 trillion kilometers (5.9 trillion miles).

matter: All things that are made of atoms or the particles that make up atoms.

million: 1,000,000; 10^6

mitochondrion: An organelle within the cells of all eukaryotes (plants, animals, etc.) that supplies energy to the cell. (plural: mitochondria)

multiverse: It has been proposed that the universe could be one of many universes that are all part of a multiverse.

naturalist: Another name for an atheist.

nucleotide: A molecular structure that forms a single letter in a genetic sequence. Three nucleotides make a codon or a word in DNA or RNA. There are four different nucleotides for use in DNA.

neutron: An electrically neutral particle of comparable mass to a proton. Neutrons are only stable within the nucleus of an atom.

neutrino: A subatomic particle of extremely small mass. The neutrino has no electric charge and only interacts with normal matter very rarely.

old earth creationism: The belief that God created the universe and the Earth but that the age of the universe cannot be determined from the Bible. This position leaves the determination of the age of the Earth and universe to science.

omnipotence: Having unlimited power. (Especially in describing God.)

organelle: Any of several different subcomponents existing within eukaryote cells that have a surrounding membrane.

prokaryote: A single-celled organism that lacks a nucleus, storing its DNA loosely within the cell. Prokaryotes are smaller and have much less DNA than eukaryotes. Bacteria are a common example.

photon: A single particle of light.

phylum: In the classification of organisms, the phylum is the highest level group within the animal kingdom. (Plural: phyla)

protein: Proteins are the basic building blocks of life. They are long chains of amino acids constructed in a predefined sequence. There are only twenty different amino acids used for building proteins, but it takes thousands of these amino acids to build a single protein. Their are thousands of different proteins in an organism, each having a unique function.

proton: A positively charged particle found in the nucleus of all atoms.

photosynthesis: The mechanism by which plants and some other life-forms use sunlight to convert carbon dioxide and water into glucose and oxygen.

quantum mechanics: The theory that describes phenomena occurring at very small scales including the wave nature of particles, the unpredictable aspects of quantum events, and the minimum size of some quantities.

quark: Quarks are fundamental particles that come in many varieties and are constituent particles of all baryons(like protons and neutrons) and all mesons. Each type of quark has its own anti-particle, which are anti-quarks.

RNA (ribonucleic acid): Single stranded genetic sequences that have been copied from DNA. RNA is used as a blueprint in the construction of proteins and is also used as a component of some biological structures and devices.

singularity: A singularity is an object that occupies only a single point and, as such, has zero volume.

solar system: A solar system consists of at least one star and all the objects orbiting it, such as planets. In our solar system the sun is orbited by eight planets, several dwarf planets, and many asteroids and comets. If two or more stars are gravitationally bound, so that they orbit a mutual center, then they are part of the same solar system, as well as any satellites they may have.

space-time: According to general relativity, space and time are not absolute universal properties but are local and can be stretched or bent under many situations. The term Space-time refers to the underlying structure to which the dimensions of time and space are mapped.

star: The closest star to earth is the sun. Stars are thousands to millions of times more massive than the earth and shine brightly due to the energy released by the fusion of hydrogen or other elements.

strata: The layering of rock built up over time by sedimentation and other processes allowing geological history to be recorded. Fossils become trapped in these layers giving an indication to the time from which they lived.

supernova: The violent explosion of a massive star at the end of its life.

theist: Someone who believes that God exists and has revealed himself to man.

theology: The study of God's nature and his relationship to creation.

thermodynamics: The set of physical laws that govern the flow of energy through a system.

topology: The study of four-dimensional structures that can describe curved space.

trillion: 1,000,000,000,000; 10^{12}

universe: The entirety of space and its contents.

visible universe: The part of the universe that could potentially be observed by telescopes. Everywhere within the maximum distance that light could have traveled to us since the beginning of time.

world view: The philosophy by which someone understands the world around them. You use your world view to fill in for unknown or unobserved aspects of reality.

young earth creationism: The belief that the Earth and the universe were created by God within the last 10,000 years.

References

General References:

1. Francis Collins, *The Language of God* (New York: Free Press, 2006)

2. Michael Corey, *The God Hypothesis* (Lanham: Rowman & Littlefield, 2001)

3. Guillermo Gonzalez and Jay Richards, *The Privileged Planet* (Washington: Regnery, 2004)

4. Peter Ward and Donald Brownlee, *Rare Earth* (New York: Copernicus Books, 2000)

5. John Sailhamer, *Genesis Unbound* (Sisters: Multnomah Books, 1996)

6. Stephen Hawking, *A Brief History of Time* (New York: Bantam Books, 1988)

7. Paul Davies, *The Goldilocks Enigma* (New York: Mariner Books, 2008)

8. Paul Davies, *The Accidental Universe* (Cambridge, Cambridge University Press, 1982)

Introduction:

1. Joe Sachs, "Aristotle: Metaphysics," Internet Encyclopedia of Philosophy, http://www.iep.utm.edu/aris-met/. Retrieved January 25, 2013.

Chapter One: The State of the Universe

1. NASA, "Sizes of Planet Candidates," (November 2013), http://www.nasa.gov/content/size-of-planet-candidates-november-2013/. Retrieved April 12, 2014.

2. Natalie M. Batalha et al., "Planetary Candidates Observed by Kepler III: Analysis of the First 16 Months of Data," *The Astrophysical Journal Supplement*, vol. 204, issue 2, pp. 37.

3. WMAP Science Team, "Wilkinson Microwave Anisotropy Probe," NASA Website, http://map.gsfc.nasa.gov/universe/WMAP_Universe.pdf. Retrieved May 18, 2010. pp. 32

4. George O. Abell, David Morrison, and Sidney C. Wolff, *Exploration of the Universe* (Philadelphia: Saunders College Publishing, 1991), pp 540.

5. WMAP Science Team, "Wilkinson Microwave Anisotropy Probe", pp. 34.

6. Ibid., pp. 32.

7. Ibid., pp 2.

8. George O. Abell, David Morrison, and Sidney C. Wolff, *Exploration of the Universe* (Philadelphia: Saunders College Publishing, 1991), pp. 605.

9. WMAP Science Team, "Wilkinson Microwave Anisotropy Probe", pp. 5.

10. Ibid., pp 20.

11. Planck Collaboration, "Planck 2013 Results. I. Overview of Products and Scientific Results," posted to arXiv (March 21, 2013), http://arXiv.org/pdf/0674450, pp. 36.

12. WMAP Science Team, "Wilkinson Microwave Anisotropy Probe", pp. 8.

13. Ibid., pp. 8.

14. Ibid., pp. 9.

15. Planck Collaboration, "Planck 2013 Results. I. Overview of Products and Scientific Results,", pp. 36.

16. WMAP Science Team, "Wilkinson Microwave Anisotropy Probe", pp. 15.

17. Ibid., pp. 15.

18. Paul Davies, *The Accidental Universe* (Cambridge: Cambridge University Press, 1982), pp. 33.

19. Paul Davies, *The Goldilocks Enigma* (New York: First Mariner Books, 2008), pp. 143.

20. WMAP Science Team, "Wilkinson Microwave Anisotropy Probe", pp. 7.

21. Ibid., pp. 32.

22. Ibid., pp. 28.

23. George O. Abell, *Exploration of the Universe*, pp.510.

24. "Abundance of the Chemical Elements," *Wikipedia*, http//:en.wikipedia.org/wiki/Abundance_of_the_chemical_elements. Retrieved February 26, 2011.

Chapter Two: Fate of the Universe

1. "Abundance of the Chemical Elements," *Wikipedia*, http://en.wikipedia.org/wiki/Abundance_of_the_chemical_elements. Retrieved February 26, 2011.

2. Fred C. Adams and Gregory Laughlin, "A Dying Universe: The Long Term Fate and Evolution of Astrophysical Objects," *Reviews of Modern Physics* (1997), vol. 69, issue 2, pp. 342.

3. George O. Abell, David Morrison, and Sidney C. Wolff, *Explo-*

ration of the Universe (Philadelphia: Saunders College Publishing, 1991), pp. 467, 511, 513.

4. WMAP Science Team, "Wilkinson Microwave Anisotropy Probe," NASA Website, http://map.gsfc.nasa.gov/universe/WMAP_Universe.pdf. Retrieved May 18, 2010. pp. 23.

5. Fred C. Adams, "A Dying Universe: The Long Term Fate and Evolution of Astrophysical Objects," pp. 345.

6. WMAP Science Team, "Wilkinson Microwave Anisotropy Probe," pp. 26.

7. Ibid., pp. 26.

8. Ibid., pp. 26.

9. S. Perlmutter, G. Aldering, G. Goldhaber, R. A. Knop, P. Nugent, P.G. Castro, S. Deustua, S. Fabbro, A. Goobar, D. E. Groom, I. M. Hook, A. G. Kim, M. Y. Kim, J. C. Lee, N. J. Nunes, R. Pain, C. R. Pennypacker, R. Quimby, "Measurements of Ω and Λ from 42 High-Redshift Supernovae," *The Astrophysical Journal* (1998), vol. 516, no. 2.

10. "Comparison of Dark Energy Models," *Space Daily* (Dec 6, 2010)

11. WMAP Science Team, "Wilkinson Microwave Anisotropy Probe," pp. 30.

12. David Sobral, Ian Smail, Philip N. Best, James E. Geach, Yuichi Matsuda, John P. Stott, Michele Cirasuolo, and Jaron Kurk, "A Large Hα survey at z=2.23, 1.47, 0.84 & 0.40: the 11 Gyr evolution of Star-Forming Galaxies from HiZELS," *Monthly Notices of the Royal Astronomical Society* (January 11, 2013), vol. 428, no. 2, pp. 1143.

13. Fred C. Adams, "A Dying Universe: The Long Term Fate and Evolution of Astrophysical Objects," pp. 345.

14. Ibid., pp. 342.

Chapter Three: The Fallacy of the Modern Assumption

1. S. Perlmutter, G. Aldering, G. Goldhaber, R. A. Knop, P. Nugent, P.G. Castro, S. Deustua, S. Fabbro, A. Goobar, D. E. Groom, I. M. Hook, A. G. Kim, M. Y. Kim, J. C. Lee, N. J. Nunes, R. Pain, C. R. Pennypacker, R. Quimby, "Measurements of Ω and Λ from 42 High-Redshift Supernovae," *The Astrophysical Journal* (1998), vol. 516, no. 2.

Chapter Four: The Fine-Tuned Universe

1. Paul Davies, *The Accidental Universe* (Cambridge: Cambridge University Press, 1982), pp. 90.

2. Paul Davies, *The Goldilocks Enigma* (New York: First Mariner Books, 2008), pp. 149.

3. S. Perlmutter, G. Aldering, G. Goldhaber, R. A. Knop, P. Nugent, P.G. Castro, S. Deustua, S. Fabbro, A. Goobar, D. E. Groom, I. M. Hook, A. G. Kim, M. Y. Kim, J. C. Lee, N. J. Nunes, R. Pain, C. R. Pennypacker, R. Quimby, "Measurements of Ω and Λ from 42 High-Redshift Supernovae," *The Astrophysical Journal* (1998), vol. 516, no. 2.

4. Paul Davies, *The Accidental Universe*, pp. 70-71.

5. Ibid., pp. 69.

6. Ibid., pp. 70.

7. "Abundance of the Chemical Elements," *Wikipedia*, http//:en.wikipedia.org/wiki/Abundance_of_the_chemical_elements. Retrieved February 26, 2011.

8. Paul Davies, *The Accidental Universe*, pp. 62.

9. Ibid., pp. 64.

10. Ibid., pp 65.

11. "Abundance of the Chemical Elements," *Wikipedia*, http//:en.wikipedia.org/wiki/Abundance_of_the_chemical_elements.

12. Evgeny Epelbaum, Hermann Krebs, Dean Lee, and Ulf-G. Meißner, "Ab Initio alculation of the Hoyle State," *Physical Review Letters* (May 13, 2011), vol. 106, issue 19, pp. 192503.

13. M. Chernykh, H. Feldmeier, T. eff, P. von Neumann-Cosel, and A. Richter, "On the Structure of the Hoyle State in ^{12}C," *Physical Review Letters* (January 19, 2007), vol. 98, issue 3, pp 32501.

14. H. Oberhummer, A. Csótó, and H. Schlattl, "Stellar Production Rates of Carbon and Its Abundance in the Universe," *Science* (July 7, 2000), vol. 289, pp. 89.

15. Ibid, pp. 89.

Chapter Five: Multiverse Theories

1. Max Tegmark, "Parallel Universes," posted to arXiv (Feb 7, 2003), http://arXiv.org/abs/astro-ph/0302131v1.

2. Ibid., pp. 4.

3. WMAP Science Team, "Wilkinson Microwave Anisotropy Probe," NASA Website, http://map.gsfc.nasa.gov/universe/WMAP_Universe.pdf. Retrieved May 18, 2010. pp. 17.

4. Max Tegmark, "Parallel Universes," pp. 5.

5. WMAP Science Team, "Wilkinson Microwave Anisotropy Probe," pp. 15.

6. Ibid., pp. 9.

7. Max Tegmark, "Parallel Universes," pp. 5.

8. Ibid., pp. 7.

9. Alan H. Guth, "Was Cosmic Inflation the 'Bang' of the Big Bang," *Beam Line* (Fall 1997), vol. 27, no. 3, pp. 18.

10. Max Tegmark, "Parallel Universes," pp. 9

11. Ibid., pp. 13.

12. "An Elegant Multiverse? Professor Brian Greene Considers the

Possibilities," *The Record* (March 21, 2011)

13. The LHCb Collaboration, "First Evidence for the Decay B^0_s --> $\mu^+ \mu^-$," *Physical Review Letters* (January 7, 2013), vol. 110, issue 2, pp. 21801-21809.

14. M. Shifman, "Reflections and Impressionistic Portrait at the Conference Frontiers Beyond the Standard Model, FTPI, Oct. 2012", posted to arXiv (October 31, 2012), http://arXiv.org/abs/1211.0004v1.

15. WMAP Science Team, "Wilkinson Microwave Anisotropy Probe," pp. 4

16. Alan H. Guth, "Eternal Inflation and Its Implications," posted to arXiv (February 22, 2007), http://arXiv.org/abs/hep-th/0702178v1. pp. 4.

17. A. V. Ivanchik, A. Y. Potekhin, and D. A. Varshalovich, "The Fine-Structure Constant: A New Observational Limit on Its Cosmological Variation and Some Theoretical Consequences," *Astronomy and Astrophysics* (1999), vol. 343, pp. 439-445.

18. Julija Bagdonaite, Paul Jansen, Christian Henkel, Hendrick L. Bethlem, Karl M. Menten, and Wim Ubachs, "A Stringent Limit on a Drifting Proton-to-Electron Mass Ratio from Alcohol in the Early Universe," *Science Express* (December 13, 2012)

19. J. K. Webb, J. A. King, M. T. Murphy, V. V. Flambaum, R. F. Carswell, and M. B. Bainbridge, "Evidence for Spacial Variation of the Fine Structure Constant," posted to arXiv (August 23, 2010), http://arXiv.org/abs/1008.3907v1. pp. 1.

20. Julija Bagdonaite, Paul Jansen, Christian Henkel, Hendrick L. Bethlem, Karl M. Menten, and Wim Ubachs, "A Stringent Limit on a Drifting Proton-to-Electron Mass Ratio from Alcohol in the Early Universe," *Science Express* (December 13, 2012)

21. Alan H. Guth, "Was Cosmic Inflation the 'Bang' of the Big Bang," pp. 19.

22. Alan H. Guth, "Eternal Inflation and Its Implications," pp. 14.

Chapter Six: Quantum Mechanics

1. Kenneth Krane, *Modern Physics* (New York: John Wiley & Sons, 1983), pp. 96.

2. Ibid., pp. 97.

3. Andreas Albrecht and Daniel Philips, "Origin of Probabilities and Their Application to the Multiverse," posted to arXiv (December 5, 2012), http://arXiv.org/abs/1212.0953v1.

4. Charles H. Bennett, "Demons, Engines, and the Second Law," *Scientific American* (November 1987), vol. 256, no. 5, pp. 108.

5. Ibid., pp. 116.

6. Zeeya Merali, "Demonic Device Converts Information to Energy," *Nature* (November 14, 2010)

7. Stuart Hameroff and Roger Penrose, "Consciousness in the Universe," Physical Letters Review (2013), http://dx.doi.org/10.1016/j.plrev.2013.08.002. pp. 1.

8. Ibid., pp. 16.

Chapter Seven: The Universal Design

1. Guillermo Gonzalez and Jay W. Richards, *The Privileged Planet* (Washington: Regnery Publishing, 2004), pp. 218.

2. Ibid., pp. 9.

3. Ibid., pp. 15-16.

4. Ibid., pp. 66.

5. Max Tegmark, "Parallel Universes," posted to arXiv (Feb 7, 2003), http://arXiv.org/abs/astro-ph/0302131v1.

Chapter Eight: Natural History of the Planet Earth

1. WMAP Science Team, "Wilkinson Microwave Anisotropy Probe," NASA Website, http://map.gsfc.nasa.gov/universe/WMAP_Universe.pdf. Retrieved May 18, 2010. pp. 28.

2. Ibid., pp. 28.

3. George O. Abell, David Morrison, and Sidney C. Wolff, *Exploration of the Universe* (Philadelphia: Saunders College Publishing, 1991), pp. 470.

4. Ibid., pp. 540.

5. Paul J. Green, "Star," *World Book Online Reference Center* (2005), http://www.worldbookonline.com/wb/Article?id=ar529540. Retrieved September 23, 2009.

6. George O. Abell, David Morrison, and Sidney C. Wolff, *Exploration of the Universe*, pp. 359.

7. Ibid., pp. 357.

8. John W. Valley, William H. Peck, Elizabeth M. King, and Simon A. Wilde, "A Cool Early Earth," *Geology* (April 2002), pp. 353.

9. Hidenori Genda and Masahiro Ikoma, "Origin of the Ocean on the Earth: Early Evolution of Water D/H in a Hydrogen-Rich Atmosphere," *Icarus* (March 2008), vol. 194, issue 1, pp. 42-52.

10. Feng Tian, Owen B. Toon, Alexander A. Pavlov, and H. De Sterck, "A Hydrogen-Rich Early Earth Atmosphere," *Science* (May 13, 2005), vol. 308, no. 5724, pp. 1015.

11. George O. Abell, *Exploration of the Universe*, pp. 360.

12. John W. Valley, "A Cool Early Earth," pp. 353.

13. John W. Valley, "The Origin of Habitats," *Geology* (November 2008), vol. 36, no. 11, pp. 912.

14. John W. Valley, "A Cool Early Earth," pp. 352.

15. Ibid., pp. 353.

16. Ibid., pp. 353.

17. Peter D. Ward and Donald Brownlee, *Rare Earth* (New York: Copernicus Books, 2004), pp. 53.

18. James F. Kasting and Shuhui Ono, "Paleoclimates: The First Two Billion Years," *Philosophical Transactions of the Royal Society*

B: Biological Sciences (June 29, 2006), vol. 361, no. 1470, pp. 920.

19. Feng Tian, Owen B. Toon, Alexander A. Pavlov, and H. De Sterck, "A Hydrogen-Rich Early Earth Atmosphere," *Science* (May 13, 2005), vol. 308, no. 5724, pp. 1015.

20. James F. Kasting, "Paleoclimates: The First Two Billion Years," pp. 921.

21. Feng Tian, "A Hydrogen-Rich Early Earth Atmosphere."

22. James F. Kasting, "Paleoclimates: The First Two Billion Years," pp. 924.

23. Feng Tian, "A Hydrogen-Rich Early Earth Atmosphere," pp. 1015.

24. Earth is 1.38 times as far from the sun as Venus and the sun only had 70% of its current solar intensity.

25. Kenneth D. Collerson and Balz S. Kamber, "Evolution of the Continents and the Atmosphere Inferred from Th-U-Nb Systematics of the Depleted Mantle," *Science* (March 5, 1999), vol. 283, no. 5407, pp. 1519-1522.

26. Ibid.

27. John W. Valley, "A Cool Early Earth," pp. 353.

28. Ibid., pp. 351.

29. James F. Kasting, "Paleoclimates: The First Two Billion Years," pp. 918.

30. Ibid., pp. 918.

31. Robert E. Kopp, Joseph L. Kirschvink, Isaac A. Hilburn, and Cody Z. Nash, "The Paleoproterozoic Snowball Earth: A Climate Disaster Triggered by the Evolution of Oxygenic Photosynthesis," *PNAS* (August 9, 2005), vol. 102, no. 32, pp. 11131.

32. Kenneth D. Collerson and Balz S. Kamber, "Evolution of the Continents and the Atmosphere Inferred from Th-U-Nb Systematics of the Depleted Mantle," *Science* (March 5, 1999), vol. 283, no. 5407, pp. 1519-1522.

33. Ibid.

34. Ibid.

35. Robert E. Kopp, "The Paleoproterozoic Snowball Earth: A Climate Disaster Triggered by the Evolution of Oxygenic Photosynthesis," pp. 11133.

36. Heinrich D. Holland, "The Oxygenation of the Atmosphere and Oceans," *Philosophical Transactions of the Royal Society B: Biological Sciences* (June 29, 2006), vol 361, no. 1470, pp. 903.

37. Ibid., pp. 907.

38. Ibid., pp. 903.

39. Ibid., pp. 909.

40. Robert E. Kopp, "The Paleoproterozoic Snowball Earth: A Climate Disaster Triggered by the Evolution of Oxygenic Photosynthesis," pp. 11133.

41. Gregory J. Retallack, Evelyn S. Krull, Glenn D. Thackray, and Dula Parkinson, "Problematic urn-shaped fossils from a Paleoproterozoic (2.2 Ga) Paleosol in South Africa," *Precambrian Research* (September 2013), vol. 235, pp. 71-87.

42. Heinrich D. Holland, "The Oxygenation of the Atmosphere and Oceans," pp. 908.

Chapter Nine: Life history on Earth

1. "Divisions of Geologic Time," *The U.S. Geological Survey Geologic Names Committee* (2006).

2. Heinrich D. Holland, "The Oxygenation of the Atmosphere and Oceans," *Philosophical Transactions of the Royal Society B: Biological Sciences* (June 29, 2006), vol 361, no. 1470, pp. 908.

3. Ibid., pp. 912.

4. Xunlai Yuan, Shuhai Xiao, and T. N. Taylor, "Lichen-Like Symbiosis 600 Million Years Ago," *Science* (May 13, 2005), vol 308, pp. 1018.

5. Linda E. Graham, Lee W. Wilcox, Martha E. Cook, and Patricia G. Gensel, "Resistant Tissues of Modern Marchantioid Liverworts Resemble Enigmatic Early Paleozoic Microfossils," *PNAS* (July 27, 2004), vol. 101, no. 30, pp. 11029.

6. Gregory J. Retallack, Evelyn S. Krull, Glenn D. Thackray, and Dula Parkinson, "Problematic urn-shaped fossils from a Paleoproterozoic (2.2 Ga) Paleosol in South Africa," *Precambrian Research* (September 2013), vol. 235, pp. 71-87.

7. Dorling Kindersley Limited, *Prehistoric Life* (New York: DK Publishing, 2009), pp. 59.

8. Daniel S. Heckman, David M. Geiser, Brooke R. Eidell, Rebecca L. Stauffer, Natalie L. Kardos, and S. Blair Hedges, "Molecular Evidence for the Early Colinization of Land by Fungi and Plants," *Science* (August 10, 2001), vol 293, no. 5532, pp. 1131.

9. Ibid., pp. 1131.

10. Heinrich D. Holland, "The Oxygenation of the Atmosphere and Oceans," pp. 908.

11. "Divisions of Geologic Time," *The U.S. Geological Survey Geologic Names Committee* (2006).

12. Dorling Kindersley Limited, *Prehistoric Life*, pp. 61-63.

13. "Divisions of Geologic Time," *The U.S. Geological Survey Geologic Names Committee* (2006).

14. Dorling Kindersley Limited, *Prehistoric Life*, pp. 30.

15. Derek E. G. Briggs and Richard A. Fortey, "Wonderful Strife: Systematics, Stem Groups, and the Phylogenetic Sygnal of the Cambrian Radiation," *Paleobiology* (2005), vol. 31, no. 2, pp. 100.

16. Dorling Kindersley Limited, *Prehistoric Life*, pp. 78.

17. Simon Conway Morris, "Darwin's Dilemma: the Realities of the Cambrian 'Explosion'," *Philosophical Transactions of the Royal Society B: Biological Sciences* (June 29, 2006), vol. 361, no. 1470, pp. 1069.

18. Dorling Kindersley Limited, *Prehistoric Life*, pp. 74-78.

19. Ibid., pp. 82-83.

20. "Divisions of Geologic Time," *The U.S. Geological Survey Geologic Names Committee* (2006).

21. Dorling Kindersley Limited, *Prehistoric Life*, pp. 96-97.

22. Ibid., pp. 96-97.

23. "Divisions of Geologic Time," *The U.S. Geological Survey Geologic Names Committee* (2006).

24. Dorling Kindersley Limited, *Prehistoric Life*, pp. 120.

25. Ibid., pp. 111.

26. Ibid., pp. 142-143.

27. "Divisions of Geologic Time," *The U.S. Geological Survey Geologic Names Committee* (2006).

28. Dorling Kindersley Limited, *Prehistoric Life*, pp. 142-144.

29. Ibid., pp. 172-173.

30. "Divisions of Geologic Time," *The U.S. Geological Survey Geologic Names Committee* (2006).

31. Dorling Kindersley Limited, *Prehistoric Life*, pp. 173, 182.

32. Ibid., pp. 172-173.

33. Ibid., pp. 196-197.

34. Ibid., pp. 206.

35. "Divisions of Geologic Time," *The U.S. Geological Survey Geologic Names Committee* (2006).

36. Dorling Kindersley Limited, *Prehistoric Life*, pp. 224.

37. Ibid., pp. 264, 323, 331.

38. Ibid., pp. 282-283.

39. Maureen A. O'leary et al., "The Placental Mammal Ancestor and the Post-K-Pg Radiation of Placentals," *Science* (February 8, 2013), vol. 339, pp. 662-667.

40. Dorling Kindersley Limited, *Prehistoric Life*, pp. 360.

41. Ibid., pp. 360-361.

42. "Divisions of Geologic Time," *The U.S. Geological Survey Geologic Names Committee* (2006).

43. Ibid.

44. Dorling Kindersley Limited, *Prehistoric Life*, pp. 388-389.

45. Ibid., pp. 454.

46. Ibid., pp. 445.

47. Ibid., pp. 461.

48. Ibid., pp. 474.

49. James Usher, *The Annals of the World* (published in Latin 1650, translated into English in 1658.)

50. Thomas Nelson, Inc., *The Nelson Study Bible, NKJV* (Nashville: Thomas Nelson Publishers, 1997), pp. 1.

Chapter Ten: Old Earth Creationism

1. Henry H. Halley, *Halley's Bible Handbook* (Grand Rapids, Michigan: Zondervan, 1965), pp. 60.

2. Thomas Nelson, Inc., *The Nelson Study Bible, NKJV* (Nashville: Thomas Nelson Publishers, 1997), pp. 4.

3. Henry H. Halley, *Halley's Bible Handbook*, pp. 59.

4. Thomas Nelson, Inc., *The Nelson Study Bible, NKJV*, pp. 4.

5. Henry H. Halley, *Halley's Bible Handbook*, pp. 61.

6. Ibid., pp. 60.

7. Ibid., pp. 60.

8. Ibid., pp. 61.

9. Maureen A. O'leary et al., "The Placental Mammal Ancestor and the Post-K-Pg Radiation of Placentals," *Science* (February 8, 2013), vol. 339, pp. 662-667.

Chapter Eleven: Origin of Species

1. Charles Darwin, *The Origin of Species by Means of Natural Selection* (First Edition 1859), ch. 13.

2. Ibid., ch. 1.

3. Ibid., ch. 10.

4. John Struthers, "On the Bones, Articulations, and Muscles of the Rudimentary Hind-Limb of the Greenland Right-Whale," *Journal of Anatomy and Physiology* (January 1881), vol 15 (part 2).

5. Charles Darwin, *The Origin of Species by Means of Natural Selection*, ch. 14.

6. Ibid., ch. 3.

7. Neil A. Campbell, *Biology* (Menlo Park, California: Benjamin/Cummings, 1996), pp. 301.

8. Ibid., pp. 286.

9. Ibid., pp. 301.

10. Ibid., pp. 1051.

11. Ibid., pp. 303.

12. Ibid., pp. 309-310.

13. Elizabeth Pennisi, "ENCODE Project Writes Eulogy for Junk DNA," *Science* (September 7, 2012), vol. 337, pp. 1159-1161.

14. Ibid.

15. Leslie A. Pray, "Transposons: The Jumping Genes," *Nature Education* (2008) vol. 1, no. 1.

16. Francis S. Collins, *The Language of God* (New York: Free Press, 2006), pp. 135.

17. Leslie A. Pray, "Transposons: The Jumping Genes,"

18. Francis S. Collins, *The Language of God*, pp. 137.

19. Ibid., pp. 131.

20. Simon Conway Morris, "Darwin's Dilemma: the Realities of the Cambrian 'Explosion'," *Philosophical Transactions of the Royal Society B: Biological Sciences* (June 29, 2006), vol. 361, no. 1470, pp. 1069.

21. Derek E. G. Briggs and Richard A. Fortey, "Wonderful Strife: Systematics, Stem Groups, and the Phylogenetic Sygnal of the Cambrian Radiation," *Paleobiology* (2005), vol. 31, no. 2, pp. 100.

22. Charles Darwin, *The Origin of Species by Means of Natural Selection*, ch. 6.

23. W. Ford Doolittle, "Uprooting the Tree of Life," *Scientific American* (February 2000), pp. 93-94.

24. Ibid., pp. 94.

25. Peter D. Ward and Donald Brownlee, *Rare Earth* (New York: Copernicus Books, 2004), pp. 84.

26. Gallop Poll: "Evolution, Creationism, Intelligent Design," (2010), http://www.gallop.com/poll/21814/Evolution-Creationism-Intelligent-Design.aspx. Retrieved January 3, 2011.

27. Francis S. Collins, *The Language of God*, pp. 198-199.

Chapter Twelve: Origin of Life

1. Leslie E. Orgel, "The Origin of Life on Earth," April 24, 1997, pp. 3.

2. Neil A. Campbell, *Biology* (Menlo Park, California: Benjamin/Cummings, 1996), pp. 298.

3. Ibid., pp. 302-303.

4. Ibid., pp. 304-307.

5. Ibid., pp. 304-305.

6. Ibid., pp. 309.

7. Ibid.

8. Ibid., pp. 310.

9. Ibid., pp. 308.

10. Peter D. Ward and Donald Brownlee, *Rare Earth* (New York: Copernicus Books, 2004), 64.

11. Neil A. Campbell, *Biology*, pp. 494.

12. Douglas Fox, "Did Life Evolve in Ice," *Discover* (February 1, 2008), http://discovermagazine.com/2008/feb/did-life-evolve-in-ice#.UunYCOTTmSw. Retrieved September 8, 2009.

13. Leslie E. Orgel, "The Origin of Life on Earth"

14. Ibid.

15. Ibid.

16. Ibid.

17. Peter D. Ward and Donald Brownlee, *Rare Earth*, pp. 60.

18. Ibid., pp. 57.

19. Leslie E. Orgel, "The Origin of Life on Earth," pp. 3.

20. Peter D. Ward and Donald Brownlee, *Rare Earth*, pp. 99.

21. Neil A. Campbell, *Biology*, pp. 114.

22. Ibid., pp. 116-117.

23. "Energy Revolution Key to Complex Life," *Science Daily* (Oct 21, 2010), http://www.sciencedaily.com/releases/2010/10/101020131700.htm.

24. Peter D. Ward and Donald Brownlee, *Rare Earth*.

25. W. Ford Doolittle, "Uprooting the Tree of Life," pp. 92.

26. "Energy Revolution Key to Complex Life"

27. Ibid.

28. Ibid.

29. Ibid.

31. Natalia N. Ivanova, Patrick Schwientek, H. James Tripp, Christian Rinke, Amrita Pati, Marcel Huntemann, Axel Visel, Tanja Woyke, Nikos C. Kyrpides, Edward M. Rubin, "Stop Codon Reassignments in the Wild," Science (May 23, 2014), vol. 344, pp. 909-913.

Chapter Thirteen: Extraterrestrial Life and Intelligence

1. George O. Abell, David Morrison, and Sidney C. Wolff, *Exploration of the Universe* (Philadelphia: Saunders College Publishing, 1991), pp. 478.

2. Peter D. Ward and Donald Brownlee, *Rare Earth* (New York: Copernicus Books, 2004), pp. 42.

3. Ibid., pp. 19.

4. Ibid., pp. 39.

5. Ibid., pp. 22-23.

6. "New Horizons Mission Timeline," New Horizons Web Site, http://pluto.jhuapl.edu/mission/mission_timeline.php. Retrieved February 17, 2014.

Chapter Fourteen: Questions, Problems, and Misrepresentations

1. Patrick Petitjean, Raghunathan Srianand, Hum Chand, Alexander Ivanchik, Pasquier Noterdaeme, Neeraj Gupta, "Constraining Fundamental Constants of Physics with Quasar Absorbtion Line Systems," posted to arXiv (May 11, 2009), http://arXiv.org/abs/0905.1516v1.

2. J. K. Webb, J. A. King, M. T. Murphy, V. V. Flambaum, R. F. Carswell, and M. B. Bainbridge, "Evidence for Spacial Variation of the Fine Structure Constant," posted to arXiv (August 23, 2010), http://arXiv.org/abs/1008.3907v1.

3. Charles Darwin, *The Origin of Species by Means of Natural Selection* (First Edition 1859)

4. V. G. Gurzadyan and R. Penrose, "Concentric Circles in WMAP Data Provide Evidence of Violent Pre-big-bang Activity," posted to arXiv (November 18, 2010), http://arXiv.org/abs/1011.3706v1.

5. Hajian, "Are There Echoes from the Pre-Big Bang Universe? A Search for Low Variance Circles in the CMB Sky," posted to arXiv (December 9, 2010), http://arXiv.org/abs/1012.1656v1.

Chapter Fifteen: Principal World Views

1. Gallop Poll: "Evolution, Creationism, Intelligent Design," (2010), http://www.gallop.com/poll/21814/Evolution-Creationism-Intelligent-Design.aspx. Retrieved January 3, 2011.

2. James Usher, *The Annals of the World* (published in Latin 1650, translated into English in 1658)

3. Gen 2:21-23.

4. Francis S. Collins, *The Language of God* (New York: Free Press, 2006), pp. 203.

Chapter Seventeen: Overview of the Theistic Universe

1. Peter D. Ward and Donald Brownlee, *Rare Earth* (New York: Copernicus Books, 2004)

Index

CPSIA information can be obtained
at www.ICGtesting.com
Printed in the USA
FSOW02n0448270515
7337FS